Digital Environments

With the title serving as an umbrella term to distinguish simulated places from real places, *Digital Environments* signifies a shift in how we think about interactions with places and spaces, both real and simulated. The very idea of *digital environments*, though, complicates such distinctions, asking simultaneously (and perhaps reductively) as to how agents engage networks, and if there can be such distinctions between virtual and real places, between agents and networks.

For ecocritics, the term brings together two concepts that are frequently cast as oppositional: *digital* standing in for *technology/technological* and *environments* often used to represent *nature/wilderness*. Thus, reading *digital environments* as *technological nature* asks us to consider not only the relationships between technologies and natures, but the very idea that there can be such distinctions, or that there might be technological natures and natural technologies. In this way, then, the play of digital/environment exposes complexities in how we theorize both technology and nature, complexities that often result in the inevitable exclusivity and polarity between the two ideas. The real and the simulated, the technological and the natural, all unfold in flagrant and complex ways that make evident the need for framing technological theories within the gaze of ecocriticism and the need for framing ecocritical theories within technological gazes.

This collection considers the possibilities of bringing ecocritical approaches into conversation with *digital environments*. The intent is to initiate a dialogue between two areas of research often understood as disparate. This book was originally published as a special issue of *Green Letters: Studies in Ecocriticism.*

Sidney I. Dobrin is a Professor of English at the University of Florida, Gainesville, FL, USA. He is the founding director/editor of TRACE Innovation Initiative. He is the author and editor of numerous books about ecology, writing, media, and technology.

Digital Environments

Edited by
Sidney I. Dobrin

Routledge
Taylor & Francis Group
LONDON AND NEW YORK

First published 2017 by Routledge

2 Park Square, Milton Park, Abingdon, Oxfordshire OX14 4RN
52 Vanderbilt Avenue, New York, NY 10017

Routledge is an imprint of the Taylor & Francis Group, an informa business

First issued in paperback 2018

British Library Cataloguing in Publication Data
A catalogue record for this book is available from the British Library

ISBN 13: 978-1-138-20770-7 (hbk)
ISBN 13: 978-0-367-14319-0 (pbk)

Typeset in TimesNewRomanPS
by diacriTech, Chennai

Publisher's Note
The publisher accepts responsibility for any inconsistencies that may have arisen during the conversion of this book from journal articles to book chapters, namely the possible inclusion of journal terminology.

Contents

Citation Information

The chapters in this book were originally published in *Green Letters: Studies in Ecocriticism*, volume 18, issue 3 (November 2014). When citing this material, please use the original page numbering for each article, as follows:

Chapter 1
Rhetoric and recapture: theorising digital game ecologies through EA's The Sims *series*
Melissa Bianchi
Green Letters: Studies in Ecocriticism, volume 18, issue 3 (November 2014) pp. 209–220

Chapter 2
Ecocomposition: writing ecologies in digital games
Kyle Matthew Bohunicky
Green Letters: Studies in Ecocriticism, volume 18, issue 3 (November 2014) pp. 221–235

Chapter 3
Rub trees, crittercams, and GIS: the wired wilderness of Leanne Allison and Jeremy Mendes' Bear 71
Sarah Jaquette Ray
Green Letters: Studies in Ecocriticism, volume 18, issue 3 (November 2014) pp. 236–253

Chapter 4
Reading environment(s): digital humanities meets ecocriticism
Stephanie Posthumus and Stéfan Sinclair
Green Letters: Studies in Ecocriticism, volume 18, issue 3 (November 2014) pp. 254–273

Chapter 5
Signals of nature, prestidigital ecology
Andrew Hageman
Green Letters: Studies in Ecocriticism, volume 18, issue 3 (November 2014) pp. 274–285

Chapter 6

The sustainability of our digital environments: the language of the upgrade path and e-waste
Caroline Stone Short
Green Letters: Studies in Ecocriticism, volume 18, issue 3 (November 2014) pp. 286–296

For any permission-related enquiries please visit:
http://www.tandfonline.com/page/help/permissions

Notes on Contributors

Melissa Bianchi is a PhD candidate in the Department of English at the University of Florida, Gainesville, FL, USA. She specializes in media studies, rhetoric and composition, and ecocriticism, and her recent scholarship establishes the intersection between game studies and critical animal studies.

Kyle Matthew Bohunicky is a PhD candidate and an Instructor in the Department of English at the University of Florida, Gainesville, FL, USA. His dissertation theorizes and analyzes how players write with and through digital games.

Sidney I. Dobrin is a Professor of English at the University of Florida, Gainesville, FL, USA. He is the founding director/editor of TRACE Innovation Initiative. He is the author and editor of numerous books about ecology, writing, media, and technology.

Andrew Hageman is an Assistant Professor of English at Luther College, Decorah, IA, USA. His research looks at techno-culture and machines in the social imaginary, ecological narratives and images in writing and film, convergences of science and fiction in science fiction, and the history of the novel.

Stephanie Posthumus is an Assistant Professor of Languages, Literatures, and Cultures at McGill University, Montreal, Canada. Her research focuses on the representations of the nonhuman, or more-than-human, in contemporary French literature, and across European literatures and cultures.

Sarah Jaquette Ray is an Associate Professor of Environmental Studies at Humboldt State University, Arcata, CA, USA. Her research interests include environmental humanities, environmental justice, new cultural geography, and cultural studies.

Caroline Stone Short is an Instructional Design Coordinator for Shadow Health, based in Gainesville, FL, USA. She graduated with a PhD from the Department of English at the University of Florida, Gainesville, FL, USA, in 2014.

Stéfan Sinclair is an Associate Professor of Digital Humanities at McGill University, Montreal, Canada. His research focuses on the design, development, usage, and theorization of tools for the digital humanities, especially for text analysis and visualization.

Introduction: Frontier 2.0

Sidney I. Dobrin

Department of English, University of Florida, Gainesville, FL, USA

In July 2016, I stood on the bank of a beautiful river early morning casting with a light spinning outfit, a small plastic jig, with the hope of hooking and landing a rainbow trout. On the far side of the river, just beyond a stand of evergreen trees, mountains rose in the pink-hued sunrise, the uppermost reaches still capped with hints of snow. A deer came to drink from the river; an eagle flew overhead. Visually, the setting was remarkable, more than realistic, almost real. I *felt* the rod react to the strike. I fought the fish to the bank, being careful not to step from the bank into the river as I reeled, adjusting my footing carefully. When the fish flopped onto the bank, a soft tone sounded in my headset, and a digital readout appeared in front of me indicating the size, species, and point value of the fish I had just landed. The visual display prompted me: "do you want to log this fish? y/n?" With my right hand—the same hand that held the rod—I indicated "no" and then removed the headset, astounded at the depth and detail of the immersive experience.

This moment of immersion unfolded *inside* the air-conditioned, steel-and-glass Citrus County Convention Center in Orlando, Florida, where I was testing a newly released virtual reality (VR) game for the HTC Vive VR system: Fishing Planet. VR, whether for gaming or other experiences, refers to the use of digital technologies to project images, sounds, and haptic sensations to replicate realistic experiences in digital—though increasingly realistic—environments. The VR experience, or what Ken Pimentel and Kevin Teixeira have described as "immersive, interactive experience generated by a computer" (p. 122), allows the user to move about, interact with, and generally participate in activities within a digital space, most often when wearing a headset interface, like the one I'm pictured in here. The sensory immersive activity of the VR experience renders the virtual indistinct from the real, giving pause not only to question the phenomenological differences, but to theorize the rationale in continuing to make such distinctions as the digital becomes both ubiquitous and indistinguishable from the real.

This same month (July 2016) I began wandering my neighborhood, campus, and larger community with my children hunting for and capturing several species of animals that have only been known to humans since about 1995: Pokémon. Our objective was to capture as many of these creatures as possible and train them to battle one another for our amusement. The Pokémon, unlike the trout I had caught in Orlando, did not reside within a digitally constructed environment; they occupied space in the material, "real" world, though they were only visible through the lenses of devices that provide augmented views of real world. Catching a Pokémon requires excursions into the material, physical world and uses mobile digital technologies to reveal information about that space that is augmented onto/over that space. Unlike VR technologies, these augmented reality (AR) technologies enmesh the digital with the "real" to provide enhanced experiences that imbricate the digital in the real and the real in the digital, obscuring the distinction between the two.

Interestingly, the popularity of the Pokémon Go phenomenon created a catalyst for scholars in ecocriticism, ecomedia, digital humanities, digital media, game studies, popular cultures, and related fields to amplify emerging work to consider relationships between digital environments and physical/natural environments. For example, in October 2016, Jason Crider employed the Pokémon Go AR experience to argue that "While Pokémon Go attempts to facilitate a conception of nature that is somewhere between technocratic, capitalistic, and stewardship, the result is endemic, perhaps even prophetic, of the way our digital technologies interact with the environment." Crider's claim opens the door for us to acknowledge the increasing need within ecocritical work to not shy away from or instinctively vilify the digital, but the necessity to bring the ecocritical gaze to the digital. The objective of this collection is just that.

The term *Digital Environments* serves as an umbrella term. Used to distinguish simulated places from "real" places, *digital environments* signifies a shift in how we think about interactions with places and spaces, both "real" and simulated. The very idea of *digital environments*, though, complicates such distinctions, asking simultaneously (and perhaps reductively) as to how agents engage networks, and, perhaps more complexly, if there can be such distinctions between virtual and real place, between agents and networks—and what might be the value in constructing such distinctions beyond a fundamentalism or nostalgia. For ecocritics, the term brings together two concepts that are frequently cast as oppositional: *digital* standing in for *technology/technological* and *environments* often used to represent *nature/wilderness*. Thus, reading *digital environments* as *technological nature* is not that far of a stretch, asking us to consider not only the relationships between technologies and natures, but the very idea that there can be such distinctions or that there might be technological natures and natural technologies. In this way, then, the play of digital/environment exposes complexities in how we theorize both technology and nature that are often expressed in overly reductive approaches that result in the inevitable exclusivity and polarity between the two ideas. The real and the simulated, the technological and the natural unfold not necessarily in new ways under the *digital environments* umbrella, but in more flagrant and complex ways that make evident the need for framing technological theories within the gaze of ecocriticism and the need for framing ecocritical theories within technological gazes.

INTRODUCTION

This collection considers the possibilities of bringing ecocritical approaches into conversation with *digital environments*. The intent is to initiate a dialogue between two areas of research often understood as disparate. On the face of it, such a dialogue might appear counterintuitive in bringing two concepts together which are often understood as working at counter objectives. In fact, more often than not, technology writ large is cast as a primary origin of environmental crisis, the very kind of situation against which much ecocritical research works. Likewise, one may also argue that those who put faith in future technologies to solve current environmental problems are simply procrastinating real environmental work that needs to be done now. But maintaining such reductive binaries in our current cultural, economic, environmental, and technological situation is no longer feasible or realistic. For just as we might read *digital environments* to distinguish simulated places from "real" places, we should also recognize that this umbrella term is most commonly understood to identify simulated environments created through the complex networks of digital devices upon which so many of us now rely. That is, in many ways, as *digital environments* become more pervasive, more ubiquitous, they become not simulated environments but the dominant environment in which information circulates, in which we circulate. The simulated space becomes indistinguishable from "real" space in rather significant ways. Or, perhaps more accurately, what we imagine to be real space must be reimagined.

On the one hand, such a conversation demands that as ecocritics we ask as to the ramifications of digital influence upon what we have traditionally identified as "natural" environments beyond the familiar technology/nature binary. On the other hand, such conversations reveal that making such distinctions between natural environments and *digital environments* may no longer be possible, that, in fact, the only significant distinction between the two are bound in the metaphors and myths through which we have come to know the digital and the environmental. Such a realization requires ecocritical consideration since it begs questions central to ecocritical agendas. For example, questions regarding the mediation of nature/environment and the place of the human within the mediation and the environment are necessary inquiries. This is to say, then, that within a context in which the digital is a growing environment, what might studies in *digital environments* gain from ecocriticism and what might ecocriticism gain from encounters in *digital environments*?

These are, of course, necessary questions, but in many ways they only initiate the inquiry that ecocritics will have to engage to make sense of environmental concerns in a digital age—and, to be a bit cynical about it, to remain relative, as well. It is, perhaps, an understatement to say that we live at a moment when technological innovation stands as the most pervasive influence on human and nonhuman existence. And, while many ecocritics might be quick to point out the correlation, those who develop, build, and circulate new technologies (not just new digital technologies, but all technologies) often fail to see—or, at least, acknowledge—the relationship between technological acceleration, technocapitalism, and the destruction of the environment. From an ecocritical perspective, the first maxim of *digital environments* ought to be "you can't talk about technology without always already talking about environments." Or as Steven Jones repeats throughout *The Emergence of the Digital Humanities*, the digital is, and always has been, physical. This maxim is more than a maxim of material implications. Yes, technological innovation can result with direct, material effects, such as e-waste and mining degradation (among others). The same innovations inevitably result in social, economic, personal, and environmental outcomes. To speak ecocritically about *digital environments*, that is to say, is to speak about complex ecologies. To speak about *digital environments* is to simultaneously speak about all environments. To speak about technological change is to speak at the same time about all change.

In the era of a new mythology, Skynet has been live (and self-aware) for about two decades. We have entered into a time in which our inquiries are no longer about how technologies work (see Wendy Hui Kyong Chun's concept of "sourcery"), but about how we negotiate with and through technologies. These are no longer the humanist inquiries of the collective value of being human, but the posthumanist inquiry that calls to question the very idea of the human agent. For the ecocritic, posthuman inquiry opens the door to neoteric inquiry, what Greg Garrard has eloquently termed the "second frontier for ecocriticism" (p. 171). Given the historical use of *frontier*—a term of particular interest to ecocritics—Garrard's label is notably fitting. We have reached a boundary with an as-of-yet only minimally explored technological/digital wilderness ahead. There are, of course, risks involved in any such exploration, the most evident, the most dangerous, and, perhaps the most hoped for, being that what we find will change everything we thought we knew.

Any conversation that invokes a metaphor of frontier also implies a metaphor of colonialism. And, we must be upfront about this: if the digital/technological represents the second frontier of ecocriticism, then it is a frontier in which acquisition, exploitation, and management will play significant roles. Part of ecocriticism's ethical imperative is to be vocal in acknowledging these aspects of the new frontier. However, don't hear this as a call to resist the digital/technological, to defend the environment from the digital; instead hear it as a call to drive up the value of the ecocritical gaze. To resist the digital is to resist the very idea that there can be things we identify as environments. That is to say, virtual worlds are worlds of great importance. They are not alien; they are not intrusive. Rather, they are recently revealed sites to which the ecocritical gaze stands to offer insight.

The contributions to this special issue move into the second frontier of ecocriticism. They are works that turn the ecocritical gaze upon *digital environments* and virtual worlds. They bring ecocriticism to game studies, the digital humanities, and film and media studies, and implicate a range of other digital studies. These articles ask us to consider the virtual as well as the material, and the relationship between the two. By no means, though, are the pieces in this collection—or the collection itself—comprehensive or exclusive. They are a beginning. In their initiation, though, they reveal an emerging disruption in the work of ecocriticism. Traditionally, ecocriticism, like all criticisms, unfolded within an idea of the representational. That is, ecocriticism addressed texts/artifacts in terms of how they represent nature, environment, and the human relationship with nature and environment. However, virtual worlds and *digital environments* exceed the representational; they are themselves natures and environments in and with which humans and nonhumans forge relationships. They disrupt the ecocritical imperative in that their binary existence as both virtual and world ask the ecocritic to consider not merely the representational qualities of the digital environment, but to consider the very possibility of constructing and inhabiting these new worlds. That is to say, whereas ecocriticism has eagerly embraced critical work that examines new media such as film, those developments remain within the critical orbit of the representational. Bringing ecocriticism to virtual worlds and *digital environments* (or is it the other way around?) breaks from that orbit and imposes ecocriticism on a new kind of artifact, one that is simultaneously virtual and real. Such a disruption, of course, does not ask that ecocriticism open its trajectory to the real, but to rethink the very notion that there can be a distinction between the real and the written, the coded and the made. In the new frontier, it is not that *nature* and *virtual* are synonymous. It is that both are ontologically obsolete. For ecocritics, this obsolescence emerges as possibility for dynamic new ecocritical approaches, for expanding already rich (and necessary) critical agendas. The chapters in this collection begin to imagine ways that those agendas might unfold.

Turning to gaming environments, Melissa Bianchi examines the ways in which game design configures relationships between nature, technology, and culture. Bianchi turns an ecocritical lens to the popular Electronic Arts video game series *The Sims*, offering a reading that examines how the design of a video game and gaming technologies reflect deep-seeded ideologies about the interactions between humans and nature. Bianchi moves, though, beyond simply pointing out how game design and technology integrate representational strategies for depicting nature. Rather, her contribution to this issue makes a much more nuanced argument about the ways in which digital technologies provide both designers and users mechanisms that allow for more complex and diverse engagement with the very ideologics through which we establish human/nonhuman relationships. For Bianchi, digital spaces foster moves away from anthropocentric and speciesist thinking. Her position is posthumanist and is influenced by her research in both game studies and animal studies.

Likewise, in chapter 2, Kyle Bohunicky also takes up game studies; Bohunicky's approach is driven by recent thinking in writing studies. Bohunicky turns to ecocompositional theories that speculate as to the relationships between discourse and environments. He places these constructionists' ideas in conversation with new materialist turns to reclaim the real outside of discourse. Bohunicky frames this conservation within an analysis of the landscapes presented in Minecraft, an independent sandbox video game. At its core, Bohunicky's article theorizes intersections of materiality and discourse by way of a game studies' methodology that understands play as procedural discourse as a manner of examining seemingly oppositional categories such as human/animal, nature/technology, and reality/utopia.

In chapter 3, Sarah Jaquette examines the relationship between the material and discursive relationship in mediating *digital environments*. Through an examination of Leanne Allison and Jeremy Mendes' 2012 interactive documentary *Bear 71*, Ray shows how human interaction with nature requires that we imagine that humans already live in "material-semiotic worlds." Ray's contribution examines the relationship between webcams and wildlife and the ways that *Bear 71* articulates hybrid geographies. Ray's work contributes to critical ecomedia studies by showing how space is not fixed, material, or objectively knowable and how technology is central to how we perceive and participate with/in nature.

In chapter 4, Stephanie Posthumus and Stéfan Sinclair consider the disciplinary spaces of ecocriticism and the digital humanities and encourages us to enable further contact between these two discursive communities. By bringing these two academic discussions together, Posthumus and Sinclair propose a more holistic "digital environmental humanities" as a kind of inquiry with rich potential for emphasizing the role of the humanities in responding to ecological crisis while also embracing new technologies and tools. Posthumus and Sinclair bring us into digital environmental humanities by the way of understanding how we read environments and by introducing a playful digital ecology in which ecocriticism and digital humanities collaborate. They emphasize this collaboration through an analysis of Michel Houellebecq's *The Possibility of an Island* that employs *Voyant*'s text analysis tools and *Neatline*'s online application for plotting time and space in texts.

Andrew Hageman looks to pedagogical approaches in thinking about the relationships between the digital and the natural when digital texts are introduced into the curriculum. Digital texts, like the iPad application "Grass," Hagemend contends, help students "dig into the ideological roots of our different interpretations of the texts as well as the ideas of Nature and Ecology that we bring to our play." Hageman argues for the "critical benefits teaching and analyzing human encounters with interactive digital texts, especially those texts with explicit or at least readily visible ecological relevance." Hageman clearly shows how digital encounters do not serve as merely escapism or entertainment, but foster opportunities for rethinking ecological and ecocritical paradigms. For Hageman, such opportunities should be connected to pedagogies to foster ecological thinking with students in a digital age.

Caroline Stone Short examines the material consequences of the digital technologies we use. She considers what it means for a technology to become obsolete and what happens to those obsolete technologies. However, Short's take on the upgrade path and e-waste pushes our consideration of e-waste beyond the familiar approach. Short argues that the relationship between digital technologies and how scholars and users write about and use those technologies have direct impact on the technologies themselves and how the next generation of digital technologies are developed. Short identifies an ethical imperative for scholars and users that they consider the material consequences of obsolescence and upgrade as they engage technologies. For Short, such a responsibility and awareness (or lack thereof) is inseparable from the material consequences rendered by technological obsolescence, such as e-waste.

I recognize that it might sound trite to say so, but in working with these chapters, I have found an inspiring energy that drives toward a new path for ecocriticism. Given that these scholars have taken up virtual worlds and *digital environments* (two terms, by the way, that require some serious critical attention on their own) within an ecocritical framework, I am energized by the possibilities that now lay before us in terms of ecocritical investigations in digital frontiers. This energy emerges not necessarily from the subjects addressed in this collection per se, but from the authorization this collection acknowledges that ecocriticism must inhabit the digital not in the ways in which it has inhabited the textual, but in dynamic, yet-to-be discovered ways.

References

Chun, Wendy Hui Kyong. *Programmed Visions: Software and Memory*. Cambridge, MA: MIT University Press, 2011.

Crider, Jason. "*Articulating Digital Heterotopias: Pokémon Go and the Augmented Future.*" 2016 University of Florida English Graduate Organization Conference: *Futures Near and Far: Utopia, Dystopia, and Futurity,* 20 October 2016, University of Florida. Gainesville, FL: Invited Talk.

Garrard, Greg. *Ecocriticsm*. 2nd ed. London: Routledge, 2012.

Jones, Steven. *The Emergence of the Digital Humanities*. London: Routledge, 2013.

Pimentel, Ken, and Teixeira, Kevin. *Virtual Reality*. New York: McGraw-Hill, 1993.

Rhetoric and recapture: theorising digital game ecologies through EA's *The Sims* series

Melissa Bianchi

Department of English, University of Florida, Gainesville, USA

This essay offers an ecocritical approach to studying games that accounts for the relationships between nature, technology and culture as they are configured by game design. By performing a reading of rhetoric and technological recapture in Electronic Arts's (EA's) *The Sims* series, the essay untangles how game design and game technology reflect ideologies – sometimes even conflicting ones – about how humans should relate to and interact with nature. This essay also argues that advancements in digital game technology afford designers and players the option to explore and perpetuate more complex and diverse ideologies about humans and nature that move away from anthropocentric and speciesist perspectives.

Despite the current popularity of digital games research, there is a relative paucity of scholarship that addresses how games simulate natural environments, animals and our interactions with them. Recent work with games has focused largely on human culture and its interactions with technology (see, e.g., Anthropy [2012], Juul [2013], Colby, Johnson, and Colby [2013], etc.) without exploring how the medium situates these two spheres in relation to nature. This humanist approach to game studies, although important, often overlooks the ideologies that games communicate about nature's relation to technology and culture. Furthermore, this primarily humanist discourse for game studies ensures the exploitation of nature both within and beyond digital environments. As Cary Wolfe observes in *Animal Rites*, 'cultural studies situated itself squarely, if only implicitly, on what looks to me more and more like a fundamental repression that underlies most ethical and political discourse … an unexamined framework of *speciesism*' (2003, 1). Wolfe's critique of speciesism's damaging role in cultural studies discourses suggests that speciesist discourse will 'always be available for use by some humans against other humans … to continue violence against the social other of *whatever* species – or gender, or race, or class, or sexual difference' (2003, 8). Speciesism, however, is not limited to our discourse and social interactions, and as Wolfe suggests, it is also a problem in our research methods. This is particularly evident in game studies where scholars have attended to the marginalisation of human subjects within games, but ignore how game practices reduce and marginalise nature, the environment and other animal species.

Thus, this essay offers an ecocritical approach to studying games that accounts for the relationships between nature, technology and culture as they are configured by game design. By performing a reading of Electronic Arts's (EA's) *The Sims* series, I offer one way we can consider how games inextricably link nature and technology together rather than maintaining an oppositional divide between them. Looking at the games' rhetoric and technological recapture, I untangle how designer's intentions and game technology produce arguments – sometimes even conflicting ones – about how humans should relate to and interact with nature. Moreover, I argue that advancements in digital game technology afford designers and players the option to explore and perpetuate more complex and diverse ideologies about humans and nature that move away from anthropocentric and speciesist perspectives.

Many game scholars, most notably Kurt Squire (2006) and Ian Bogost (2007), have argued that the rhetoric of a digital game's semiotic elements (text, images, cinematics, etc.) and interactive processes (rules and game mechanics) communicate its ideologies. Semiotic elements and interactive processes, however, are shaped equally by game designers' intentions and the capabilities and limitations of the technology viable to them. As such, procedural rhetoric is a crucial concept for reading and critiquing games, and it provides a critical foundation for understanding nature and technology within digital games. Bogost defines procedural rhetoric as argumentation that relies on rule-based interactions as a mode of signification. He explains that in games, 'arguments are made not through the construction of words or images, but through the authorship of rules of behaviour, the construction of dynamic models' (2007, 28–29). Put another way, rules and processes mediate player interactions in digital games, and the way players negotiate this mediation creates specific arguments about players and their relations to certain signifiers within the game.

While game designers author these rules and procedures, their work is often also limited by the hardware's technological constraints. Media scholar Terry Harpold explains that '*Gameplay is the expression of combinations of definite semiotic elements in specific relations to equally definite technical elements*' (2008, 91, emphasis in the original). Harpold's definition suggests that reading ideologies in games must also account for the relations between signifying features and the technologies that influence their in-game presentation. The limits of a game's hardware and software determine, to varying degrees, the objects and rules within its world (2008, 92), which means that digital games' cultural signifiers – especially those for animals and nature as a whole – are inherently bound to and governed by material limitations (i.e., processing speed, memory, graphical capabilities, keyboard and mouse controls, etc.) and not just the cultural context of the designers and players.

Furthermore, because designers use narrative elements to recuperate hardware failure, constraints and compatibility, these technical elements play a significant and material role in representation. Harpold explains this concept as recapture. Recapture occurs

> on the cusp of a sort of crisis in representation: exactly at the moment where entanglement threatens to bring forward the game's determinism by its definite technical situation, that determinism is turned back into the gameworld, so as to seem to be another of its (arbitrary but consistent) rules. (2008, 93)

In essence, moments of recapture obscure or rescript the technical elements that make up the game world in ways that seem consistent with the game's rules or narrative. Recapture then has narrative implications for games, however, it can also impact a game's rhetoric. Despite designers' intentions for internal consistency, instances of recapture in digital games can produce procedural rhetoric that subverts the rhetoric of other elements within the game. Examining *The Sims* series allows us to see how the insertion, removal and redesign of procedural representations and recapture can change the games' arguments and ideologies about the natural environment in productive ways.

I analyse *The Sims* series in particular because of its widespread popularity and longevity in the digital games market, as well as its observable changes in game design over the last decade and a half. Each of the base games in the series – *The Sims* (Maxis 2000), *The Sims 2* (Maxis 2004) and *The Sims 3* (The Sims Studio 2009) – along with their corresponding expansion packs have sold over 150 million units internationally as of May 2011 (Sacco 2011). Because of their record-breaking sales, the game series has seen several remakes with an upcoming fourth iteration in 2014. Upgrades to the technologies (both hardware and software) available for designing and running *The Sims* have significantly allowed the games to reimagine how we conceptualise the links between humans, animals and machines. Over time, the series appears to move away from an anthropocentric understanding of these relations, in part, a result of its improved technology, which enable the series to embrace both ecocentric and posthumanist perspectives. These changes illustrate how rhetoric and recapture shape and dismantle categorical binaries like nature and technology as well as human and animal. By analysing how these categories are re-scripted across *The Sims* series, it is evident that (1) as technology improves and the need for recapture in *The Sims* games disappears, their ideologies about nature become more ecocentric and (2) technological upgrades allow for a more complex play experience that challenges speciesist and anthropocentric conceptualisations of the relationship between humans, animals and objects.

A brief overview of EA's *The Sims*

EA's *The Sims* was released on 4 February 2000 for personal computers (PCs) and since then has spawned a number of expansion packs and sequels for various game platforms. *The Sims* games are classified as life simulators because they use an agent-based artificial life program to direct the actions of simulated life forms or Sims. Programmed with an artificial intelligence system, Sims respond autonomously to environmental conditions within the game world. Using what the game calls 'free will,' Sims will attend to their 'needs' or 'motives,' which include biological factors, such as hunger and energy, as well as psychological ones, such as fun and socialising. The satisfaction of a Sim's needs is measured by several green bars that will turn red if the Sim does not perform the actions necessary for satisfying the need. The role of players in *The Sims* is to direct and manage the actions that Sims perform because their 'free will' tends to lead them astray from performing actions that effectively satisfy their needs. Satisfying a Sim's needs often has positive consequences in the game, while ignoring them often has negative results. Additionally, there are some actions that Sims will not perform without player input to do so, such as conceiving children, finding a job and paying bills.

Beyond the games' simulated life forms, several aspects of gameplay have remained relatively the same across *The Sims*, *The Sims 2*, *The Sims 3* and their expansions packs. For example, different game modes pose particular constraints on the actions that players can take within the game world. In Live mode, players can control the passage of game time as well as instruct their Sims to interact with objects or other Sims. In contrast, Buy mode and Build mode automatically pause game time so that players can purchase and arrange furniture in their Sims' house or construct and renovate their home or neighbourhood. Additionally, each of the base games contains a Create-a-Sim mode in which players build and customise the physical appearance and personality of their Sims. Each of the iterations of *The Sims* game has added to the number of customisation options for each of these modes of play. How players interact with and use the game's artificial intelligence, architecture systems and play modes remains largely undirected by the game itself. This design promotes open-ended gameplay with no clear way of winning or losing.

In describing some of the major aspects of *The Sims* series, the anthropocentric design of its in-game procedures becomes evident. Most obviously, this is because the game is targeted at human players, but it is also because human Sims are the primary focus of the play experience and its processes. In 2002, however, EA released an expansion pack for *The Sims* entitled *The Sims: Unleashed* (Maxis 2002), which integrated animal Sims into human Sims' households. The popularity of *Unleashed* among the series' fans spawned similar expansion packs for the second and third iterations of the game, which were simply entitled *Pets*. How these expansion packs alter the gameplay of the original games and how they too have changed over time demonstrates that our changing conceptions of the links between humans, animals and technology are not independent of one another, but rather shaped by an ecological and co-evolutionary process. The *Pet* expansion packs push towards a contentious perspective of nature that destabilises players' anthropocentric perspectives through ecocentric and posthumanist representations of a digital ecology.

Navigating the borders between neighbourhoods and nature

In *The Idea of Wilderness*, Max Oelshlaeger outlines several conceptions of nature that we can use to read the design of 'nature' in *The Sims*. Oelshlaeger argues that contemporary wilderness philosophy is fractured among several different ideologies, which include resourcism, preservationism, ecocentrism, deep ecology and ecofeminisim. Each of these ideologies offers a unique ethical perspective of how humans should interact with the natural world. *The Sims* and similar simulation games perpetuate these ideologies through the ways that procedural rhetoric and recapture shape their rules and determine how humans (both players and Sims) interact with representations of wilderness. Moreover, as game studies scholar Miguel Sicart (2009) argues, games are themselves designed ethical systems and experiences, and their rules have moral values that affect players. This understanding of games as 'ethical objects' and also as digital environments makes them uniquely suited for simulating the ethical systems Oelshlaeger outlines in his work. Using *The Sims* games, I demonstrate how the values of resourcism, preservationism and ecocentrism are implemented in the games' designs as technological upgrades to the game engine and its hardware facilitate more complex human–nature interactions while reducing the need for moments of spatial and environmental recapture.

The design of the neighbourhood in *The Sims* suggests that humans are estranged from, rather than a part of, nature. In the original game, players engage with the neighbourhood screen, an interface that contains 10 rectangular residential lots of varying sizes arranged along a road that loops through grassy, forested and mountain terrain. Players can click on these lots to 'enter' them and interact with the land, house and Sims. Clicking any of the surrounding greenery on the neighbourhood screen, however, does nothing. The neighbourhood's natural surroundings are procedurally 'dead,' and players cannot enter into these areas in the same way that they do residential lots. The neighbourhood's natural environment is not interactive because there are no coded processes supported by these areas of the map. Only land to be developed (the residential lots) serves a functional purpose for players, and so these spaces are highlighted when the cursor is moved over them. In part, this design choice is both a technological and a conceptual one. To make the trees and mountains on the neighbourhood screen interactive would take more processing speed and random access memory (the minimum requirements for *The Sims* are 233 MHz and 32 MB for Microsoft Windows and 233 MHz and 64 MB for Mac) or a reallocation of these resources to do so. Because the game was originally intended to simulate human domestic life, the outdoor environment of the game was not as highly prioritised, and thus, it was kept inactive and removed from human players' accessibility out of technological necessity.

For similar reasons, human Sims cannot be placed by players or interacted within the game's simulated natural environment. Any procedures and processes players choose to have their Sims perform are confined within the neighbourhood's lots. Within the lots themselves, any land, fauna or flora that lies beyond the residential domain remains greyed out and 'dead' – another design choice made to cut down on loading times and memory usage. Clicking on these grey spaces in the game world elicits no response from the interface, and like the 'dead' space of the neighbourhood screen, they visually and procedurally separate players from the game's representation of natural environments. This type of interface design reinforces what Oelschlaeger calls resourcism. He explains that resource conservationists view wild nature as 'a means only to human ends,' and as such 'humankind is understood not as a part of but apart from the green world. Nature is presumed to be merely inert matter-energy devoid of value until the humanizing force of civilization is forced upon it' (Oelschlaeger 1991, 316). The natural spaces in *The Sims* model Oelschlaeger's definition as they remain inert features of the games interface. In contrast, the interactive house lots – spaces upon which humanising forces (the player and Sims) may act – have more value as the sites where the majority of game play occurs. The game highlights their significance both visually, when the residential lots are moused-over, and procedurally, when players interact with these spaces allotted for development.

By designing a game in which simulated nature is predicated on, if not synonymous with, technology and its limitations, game designers link nature and technology together so that conceptions about and interactions with one directly influence the other. For example, many of the features that *The Sims: Unleashed* adds to the base game change how the separation between nature and technology is perceived by remapping how players interact with and control land in the game world. The expansion pack ups the minimum requirements of *The Sims* to 350 MHz processing speed and 64 MB of RAM for Microsoft Windows, and as a result, it is able to increase the size of the base game's neighbourhood from 10 lots to over 40. For the first time in the series, players are granted the ability to rezone these lots for either residential or community use. While residential lots are dedicated living spaces for Sims and their homes, community lots function as commercial hubs in which players can build shops, restaurants and other similar establishments. Players can opt to build natural parks in commercial lots and have their Sims visit and interact with these areas as pseudonatural areas. The emphasis placed on the processes of land acquisition and development in *Unleashed* illustrates how improvements in the game's technology, particularly through offering players more memory and options, in the game environment, can still coincide with conceptions of nature as a resource and yet simultaneously offer players a comparatively more interactive relationship with representations of nature.

When EA released *The Sims 2* in 2004, advancements in computer hardware allowed players even more interactions and control over the design of land in the game. As a result, the procedural rhetoric of the space advocates for a more preservationist, rather than resourcist, approach to nature. *The Sims 2* requires a PC with a minimum of 1.2 GHz processing speed and 256 MB of RAM, a substantial upgrade from the original game. With faster processing speeds and more game memory, the game allowed players the option of constructing their own neighbourhoods by choosing from pre-made templates with different road, lot and terrain layouts. Players can choose from four types of terrain: a lush, green temperate aesthetic, a sand-coloured desert one, brown-coloured dirt, or concrete. Once players have built a neighbourhood, they can click on the 'Decorations' button to add, replace or relocate flora, landmarks, weather effects, stones and other miscellaneous items in the neighbourhood spaces unoccupied by residential or community lots. For players, these design options render what was previously inert space interactive.

This interactivity with natural spaces suggests that *The Sims 2* models preservationism more closely than resourcism. Oelshlaeger argues that preservationism maintains that 'nature is a living whole in evolutionary process' and yet it 'remains anthropocentric since all values rest on the assumption that human beings are the apex of creation and the measure of all worth' (1991, 316). In effect, preservationism is concerned with protecting existing wild species, and it realises the value of nature to and its inseparability from humanity. At its core, however, nature's significance remains to be assessed from human perspectives. In the game, players cannot participate in the preservation of wild species (there are no wild species that need preserving), but they can interact with wild nature in ways that make it as significant to the processes of developing residential and commercial lots. In these instances, the value of the game's natural space is still largely determined by players' decisions about how best to design and decorate the surrounding wilderness. Thus, *The Sims 2* comes close to modelling some of the ethical aspects of preservationism, though the analogy is somewhat inexact.

In contrast, *The Sims 3* reimagines neighbourhoods as ecological spaces that link rather than divorce civilisation from nature. In the game, neighbourhoods are referred to as 'worlds,' a product of both EA's use of the Create-a-World tool and also the increased size of neighbourhoods. The change in terminology signifies the games' shift towards understanding the surrounding environment of the game space as a world system in which human life is entrenched. Unlike the neighbourhoods of *The Sims* and *The Sims 2*, worlds are 'seamless,' meaning that players are no longer limited to playing in one lot at a time with greyed out spaces where the natural environment should be. Upgrades to processing speed, RAM and graphics cards allow players to enter lots and load visual data from the surrounding areas to create an immersive environment. Although older versions of the game isolated game time and actions in each lot from the others, *The Sims 3* does away with this design, unifying the different parts of the game world. In this iteration of the game, players can see and openly explore other lots as they play, fluidly integrating movements and actions at the lot level with those occurring in the neighbourhood, such as a Sim travelling from one lot to another.

The design emphasis placed on dynamically uniting the neighbourhood lots with their surrounding wildlife reveals a shift in emphasis in the game's design from home to world. Such a refocusing of the game's design suggests an ideological move towards ecocentrism because the game's rules now place value on players' interacting with the game's environment and its inhabitants as a unified whole. To refer again to Sicart (2009), whenever players confront and experience an intentionally designed space and its rules, they are always also experiencing an ethical orientation. According to Oelschlaeger (1991), ecocentrism escapes the anthropocentric biases of resourcism and preservationism by 'inverting the relation between our species and the natural world' (316). *The Sims 3* reorients the player and Sims' relation to the 'natural' world by revamping its resourcist, and preservationist play mechanics to simulate the reality that human life is situated within a larger, ecological context that incorporates undeveloped landscapes.

Moreover, *The Sims 3: Pets* (The Sims Studio 2011) expansion pack emphasises ecocentrism in narrative structures that juxtapose an ecocentric view of nature with resourcist and preservationist perspectives. Oelschlaeger (1991) explains, 'ecocentrism embraces holism and overcomes speciesism. Yet culture refutes any thesis that the human project is strictly natural or that the good life can be equated with a natural life: to be human is to be enframed by language and history' (316–317). *The Sims 3: Pets* addresses the tension between ecocentric holism and cultural immersion through a new world called Appaloosa Plains. Aptly named after the Appaloosa breed of horse, this world is a Midwestern ranch style town oriented around the interactions between human Sims, their animal companions, and how these interactions are contextualised by natural and cultural contexts. For example, in addition to community centres like the grocery store, bistro, school and so on, Appaloosa Plains possesses an equestrian centre, cat jungle gym and several nature parks, which act as spaces where Sims can travel and interact with their pets in a variety of ways. These spaces explicitly designated for interspecies and nonhuman species interactions acknowledge and emphasise to players the importance of human–animal relations moving further away from the anthropocentric design of the neighbourhood in the first version of *The Sims*.

Although I have argued that advancements in game technologies help designers move away from anthropocentrism, this is not always the case. Technological upgrades only offer designers the opportunities to reexamine and reinvent their procedural rhetoric, not necessarily the impetus. This is most apparent if we observe the place of deep ecology and ecofeminism in *The Sims* series as both ideologies are tenuously acknowledged in *The Sims 2: Seasons* (Maxis 2007) and *The Sims 3: University Life* (The Sims Studio 2013) expansions packs. Oelschlaeger identifies deep ecology as being 'consistent with preservationism (holism) and ecocentrism (rejection of speciesism) even while going beyond these ideas of wilderness by embracing a mélange of additional ideas: bioregionalism, ecofeminism, and green politics' (1991, 317). Many of the tenets of deep ecology are embodied by what are called 'PlantSims' in the games. PlantSims are human–plant hybrids that have green skin and foliage for hair. They are created in the game when a human Sim overuses pesticides on crops or fruit trees, eats a 'Forbidden Fruit,' or participates in scientific research and experiments. When a Sim becomes a PlantSim, his or her visible motives change to the need for love, water and sunlight, and they gain the ability to talk to plants which boosts the plants' health dramatically. Deep ecology maintains that all life has intrinsic value and that the quality of that life should supersede its quantity. In effect, deep ecologists value 'green societies' and PlantSims embody this ideology both literally and figuratively. PlantSims tightly couple the human to the natural environment through their plant-like physiology, needs and interactions. Their lives are demonstrably valued as much as other Sims in that game rules and procedures indicate to players that they should interact with PlantSims as they would to other humans, guiding them to fulfil their needs. This play structure supports actions that maintain all life – regardless of its type – as having intrinsic value.

Although deep ecology is somewhat evinced by PlantSims in the games, ecofeminism is by far more difficult to observe, if it is present at all. Oelshlaeger (1991) defines ecofeminism as an 'identification of a complicated problematic involving sex-based ideology' (317) that maintains androcentrism as the root cause of 'ecosystemic malaise' (310). *The Sims 2: Seasons* approaches this idea that ecological issues are rooted in sex-based ideologies about nature through the game's first PlantSim family, the Greenmans. In this family's narrative, there is a 'naturally' occurring PlantSim gene that is passed from mother to daughter through spores, rather than through heterosexual reproduction. The family's story genders nature as expressly female; however, because the resolution of family narratives in *The Sims* are left up to the intervention and discretion of players, whether this design serves as an ecofeminist critique or not is contentious at best.

Despite the interesting and potentially progressive ideologies about nature that PlantSims represent, their place in the series is still a relatively minor one. As part of varying expansion packs, their presence in the game is optional and unnecessary for running the base game by itself. Thus, while improved technologies may afford designers and players the option to explore and perpetuate more complex and interesting ideologies about humans and their relation to nature, there is not always a correlation between the two. For the most part, however, The Sims games seem to model Oelshlaeger's taxonomy of contemporary conceptions of wilderness, progressively erasing the distinctions between technology and nature through a remapping of how humans – both those simulated in the game and those who play it – interact with nature, real or simulated.

Playing with pets

The design of nonhuman animals in *The Sims* games also illustrates how rhetoric and recapture play a role in expressing changing ideologies about nature and its place in the human home. Each new version of *The Sims* pet expansion pack erases distinctions between humans, animals and machines. This section of the essay analyses how the games gradually shift from an anthropocentric to a post-humanist perceptive by framing Sims as cyborgs. Drawing on Donna Haraway's (1991) 'A Cyborg Manifesto,' a cyborg is 'a cybernetic organism, a hybrid of machine and organism, a creature of social reality as well as a creature of fiction' (149). They are 'creatures simultaneously animal and machine, who populate worlds ambiguously natural and crafted' (Haraway 1991, 149). The hybridity of Haraway's cyborg is located in the fictional world that it inhabits, and its organismal and mechanical parts are taken from that specific environment. In keeping with Haraway's myth, simulated life forms in digital games act as cyborgs given their hybrid construction. Sims are conceptually tied to both the organic and the mechanical because their design features refer to characteristics of real animals while they are simultaneously constructed from the game's technology. Identifying Sims as cyborgs can help us understand how their construction as both organism and technology dismantles the boundaries and hierarchies between humans, animals and objects. Haraway (1991) writes:

> No longer structured by the polarity of public and private, the cyborg defines a technological polis based partly on a revolution of social relations in the oikos, the household. Nature and culture are reworked; the one can no longer be the resource for appropriation or incorporation by the other. (151)

The cyborg's characterisation as neither entirely public nor private is especially pertinent to *The Sims* because the game series explicitly began with the social relations of the home as central to its subject matter before progressing to world-building. As we have seen, later versions of the game de-centralise the home from the focus of gameplay by adding more ways for players to interact with both residential and nonresidential areas together as an ecological system. By identifying Sims as cyborgs, and especially by introducing nonhuman Sims (or cyborgs), the games allows us to consider how we may rescript the relationship between nature, culture and technology.

The first pet expansion pack of *The Sims* series suggests that there are ontological similarities between humans and other animals through the design of dog and cat Sims. Aptly titled *Unleashed*, this expansion pack allows players to add pets to their human Sims' households; however, players have nearly no control over these dogs and cats when compared to their human counterparts. At most, players can direct these digital animals to

'Go Here,' moving them to different locations within an active lot. In contrast, human Sims have a variety of actions available to them depending upon the object or other Sims the player would like them to interact with. It would seem then that dogs and cats function more like objects in the game rather than life forms. Similar to appliances and furniture in the game, it is only when a player selects a human Sim to control and interact with a pet does a list of possible actions the human Sim could take appear on the screen. For example, players can select a human Sim and then click on a dog to give it a command like 'Attack' or 'Sit & Stay,' but they can never make the dog or cat do this on its own by selecting it. And yet, unlike objects, dog and cat Sims do evince free will, perhaps even more so than human Sims, because their actions cannot be controlled by the hand of the player. Pets freely move around the lots on which they live and generate their own actions and interactions with objects, such as 'Bark at door,' 'Fetch paper' or 'Chase tail.' The free will that cats and dogs possess in *Unleashed* suggests that there is little difference between the ontology of humans and that of other animals. At the same time, players' inability to access and control cat and dog behaviours reveals that, despite sharing similar ontologies with other animals, humans can only ever imagine, never experience, what it is like to be any other being. Thus, the game is designed so that players can only interact with and exert their power over other animals through human Sims.

Other aspects of animal design in *Unleashed*, however, draw attention to the technological apparatus of the game and how it is instrumental in shaping and perpetuating what appears to be a Cartesian understanding of nonhuman animal species. Despite differences in head shape and colour, for the most part, cat and dog Sims share very similar models in the game, and these models are programmed to move through and interact with the game space in similar ways. Wild animals in the game, namely raccoons and skunks, are re-skinned versions of these domesticated animals and have an even more limited role in game play. Raccoons and skunks occasionally wander onto residential lots. When they do, they not only cannot be controlled by players, but also perform a small, scripted set of actions. Human Sims can interact with these animals, but the interactions are limited. Sims can 'shoo' raccoons from digging through their garbage (which may not always be effective) and 'pet' skunks (which may end in spraying). Although the similarities between dog, cat and human Sims illustrate the fluidity of the human–animal dialectic, wild animals in the game reinforce the division between humans and animals, as well as wild and domestic, for players through what Eileen Crist (1999) identifies as mechanomorphism or their repetitive, machine-like processes.

In a similar vein, small pets like rodents, birds, fish and reptiles function more like objects than their canine and feline counterparts. The technological limits that prevent complex animal models for these species are recaptured in the game by designing these creatures as objects in the game rather than as active Sims. Unlike cats and dogs which must be adopted through a Pet Centre or as strays that randomly appear on lots, small pets must be purchased in Buy mode or from in game stores like furniture. Similar to other objects in the game, small pets take up a certain number of floor tiles that correlate to the size of their cage or tank. Human Sims can watch, feed and play with these animals if the player clicks on the cage and selects an interaction. Small pets do not have free will like other Sims in the game, and they remain confined to the tank or cage, moving in scripted and repetitive patterns. The recapture used in designing these animals argues that smaller animals are consumer goods and not much more, a stark contrast from the close association between humans, dogs and cats. For example, players can 'restock' small pet tanks or cages in the same way that they 'restock' the refrigerator with food when these pets die. Players can have their human Sims talk to parrots in the game, earning charisma skill points, as they would if they practiced talking to themselves using a mirror. Human Sims can interact with dogs and cats in ways that are fundamentally different, in that they satisfy the motive to socialise for both parties. Small pets have no motives. The limited way small animals appear in the game and are interacted with speaks to the game's inability to support the technological processes for developing these animals further, and yet it also reveals a conflict in the game's conception of animal ontology. While some animals exist as living individuals in the game, others are little more than objects, and this division in the animal category marks the beginning of how the series complicates our understanding of human and object ontology.

With the releases of *The Sims 2: Pets* (Maxis 2006) and *The Sims 3: Pets*, dogs, cats and eventually horses in *The Sims 3* are redesigned to be more like human Sims, challenging the anthropocentricism of their antecedents. Both expansion packs add the Create-a-Pet feature to the basic game that allows players to build their own dogs and cats in a manner similar to creating human Sims by choosing both physical traits (coat patterns and colours, body shape, facial structure) and personality characteristics (playfulness, neatness, loyalty, aggression and intelligence) for their pet. Dogs and cats in this expansion pack also have the same needs as human Sims, like 'social,' 'bladder,' 'energy,' 'hunger,' 'fun,' 'comfort' and 'hygiene.' Pet Sims, however, do not require players to fulfil an 'environment' need, which for dogs is replaced by 'chew' and for cats 'scratch.' This substitution acknowledges both the alterity of these species and their similarities to humans. Similarly, pets can also hold jobs like humans, but only in three specific career paths, each with its own required skill set: security, showbiz and service. In *The Sims 2: Pets*, dogs and cats are still independent and autonomous, and though they still cannot be directly controlled by players, the system for training them has been updated. Through instruction and reinforcement, pets can learn commands much in the same way that toddlers in the game learn to walk, talk and use the toilet – another set of processes, demonstrating the similarities between humans and other animal species.

Although in *Unleashed* dog and cat Sims had very little in common with human Sims, in the *Pets* expansion packs, it is evident that game designers have given these pet simulations behaviours and traits that were once exclusive to human Sims in the game. These developments mirror the movement away from anthropocentrism that occurred concurrently with moves towards ecocentrism in the previous century. Haraway (1991) explains that

> By the late twentieth century in United States scientific culture, the boundary between human and animal is thoroughly breached. The last beachheads of uniqueness have been polluted if not turned into amusement parks – language, tool use, social behaviour, mental events, nothing really convincingly settles the separation of human and animal. And many people no longer feel the need for such a separation. (152)

While *The Sims* games were first published at the start of the twenty-first century, the games' ideologies have not quite caught up to those of the previous century. It is only through technological upgrades that the games can more accurately simulate the boundary breaches between humans and animals. Technology, then, works as a driving force to de-centre the human from both the focus of gameplay and its ideologies, pushing for a more ecological understanding of how humans are situated in relation to other animals and nature.

For example, the faster processing speeds, larger memory capacities and greater graphical capabilities in *The Sims 3: Pets* further develop the design of cats, dogs and horses, so that playing with these animals is procedurally similar to playing with human Sims. In addition to motives, the major pets in the game now also have wishes and lifetime rewards. This is a drastic change from the earlier games, which did not reveal what pets desired to do or achieve in their lifetimes. Players are made privy to the drives and desires of these animals through processes that mirror human Sims, closing the gap in how different the interactions are between human and nonhuman animals. At the same time, however, these wishes and lifetime rewards remain unique to the species of animal they belong to, avoiding the pitfalls of anthropomorphising their behaviours. Dogs will desire to 'sniff' or 'chew' on certain objects, compared to human desires, which may focus on goals related to careers or homeownership. While cats, dogs and horses can be interacted with like human Sims, 'minor pets' still function much like small pets in earlier games; however, they have diversified these animals into specifically named species possessing price values that depend upon their rarity in the game. These minor pets cannot be bought though they can be sold. To obtain a minor pet, the animal must be found in the game's natural environment, captured and brought into the home. Minor pets, however, may escape back into the wild, suggesting a fluidity to the relationship between the outdoor environment and the domestic space of the home as these animals circulate across the nature–culture divide.

These changes in the way animal Sims function reveal how game technologies and game designers' intentions work both in conjunction with and against one another to produce apparent shifts in the games' ideologies about human–animal relations. Technological limitations that force designers to implement moments of recapture tend to reinforce an anthropocentric perspective of natural environments and nonhuman animals that reduces and marginalises these entities. With life simulation games, the coupling of technical limitations with procedural rhetoric can reify real world discourses that maintain that humans are separate from nature and that we are ontologically different from other animals. It is important then that we understand how these elements are at work when we play computer games if we are to be critical players and scholars.

On a broader scale, Haraway's cyborg myth is one method we can use to analyse how digital games triangulate the links between nature, culture and technology without framing such arguments from a wholly anthropocentric or humanist perspective. Conceptualising simulated beings as cyborgs allows us to account for the impact of our ideologies about nature on both the organic (via designers, players and in-game references to living entities and natural space) and the mechanical (via in-game objects and game hardware). Haraway (1991) suggests that 'a cyborg world might be about lived social and bodily realities in which people are not afraid of their joint kinship with animals and machines, not afraid of permanently partial identities and contradictory standpoints' (155). Digital games are cyborg worlds in which we explore how player and designer interactions with nature–technology – or nature as it is simulated through technology – can lead to changes in our conceptions of how these seemingly disparate categories are inextricably linked to one another through rhetoric and recapture. By being attentive to how our ideologies about the natural environment and animals are presented in digital games, we can see some of the problematic ways we reduce and marginalise nature, as well as how these processes are reified, challenged or both by gameplay and its technological apparatus.

Acknowledgements

I thank Sidney Dobrin, Terry Harpold and Susan Hegeman for their comments, suggestions and editorial advice throughout the research and writing process. Thank you as well to Kyle Bohunicky for support and encouragement. I also thank the anonymous reviewers for their feedback.

References

Anthropy, A. 2012. *Rise of the Videogame Zinesters: How Freaks, Normals, Amateurs, Artists, Dreamers, Drop-outs, Queers, Housewives, and People Like You Are Taking Back an Art Form.* New York: Seven Stories Press.
Bogost, I. 2007. *Persuasive Games: The Expressive Power of Videogames.* Cambridge, MA: MIT Press.
Colby, R., M. S. S. Johnson, and R. S. Colby, eds. 2013. *Rhetoric/Composition/Play through Video Games: Reshaping Theory and Practice of Writing.* New York: Palgrave Macmillan.
Crist, E. 1999. *Images of Animals: Anthropomorphism and Animal Mind.* Philadelphia, PA: Temple University Press.
Haraway, D. J. 1991. *Simians, Cyborgs, and Women: The Reinvention of Nature.* New York: Routledge.
Harpold, T. 2008. "Screw the Grue: Mediality, Metalepsis, Recapture." In *Playing the Past: History and Nostalgia in Computer Games*, edited by Z. Whalen, and L. N. Taylor, 91–108. Nashville, TN: Vanderbilt University Press.
Juul, J. 2013. *The Art of Failure: An Essay on the Pain of Playing Video Games.* Cambridge, MA: MIT Press.
Maxis. 2000. *The Sims.* Redwood City, CA: Electronic Arts.
Maxis. 2002. *The Sims: Unleashed.* Redwood City, CA: Electronic Arts.
Maxis. 2004. *The Sims 2.* Redwood City, CA: Electronic Arts.
Maxis. 2006. *The Sims 2: Pets.* Redwood City: Electronic Arts.
Maxis. 2007. *The Sims 2: Seasons.* Redwood City, CA: Electronic Arts.
Oelschlaeger, M. 1991. *The Idea of Wilderness: From Prehistory to the Age of Ecology.* New Haven, CT: Yale University Press. http://sims.wikia.com/wiki/File:El_Bandito_strikes.JPG
Sacco, D. 2011. "The Sims 3: Generations." *Market for Computer and Computer Games.* McCormack and Morrison. http://www.mcvuk.com/retail-biz/recommended/read/the-sims-3-generations/02925.
Sicart, M. 2009. *The Ethics of Computer Games.* Cambridge, MA: MIT Press.
Squire, K. 2006. "From Content to Context: Videogames as Designed Experience." *Educational Researcher* 35 (8): 19–29. doi:10.3102/0013189X035008019.
The Sims Studio. 2009. *The Sims 3.* Redwood City, CA: Electronic Arts.
The Sims Studio. 2011. *The Sims 3: Pets.* Redwood City, CA: Electronic Arts.
The Sims Studio. 2013. *The Sims 3: University Life.* Redwood City, CA: Electronic Arts.
Wolfe, C. 2003. *Animal Rites: American Culture, the Discourse of Species, and Posthumanist Theory.* Chicago, IL: University of Chicago Press.

Ecocomposition: writing ecologies in digital games

Kyle Matthew Bohunicky

Department of English, University of Florida, Gainesville, USA

This paper contributes to digital writing studies and the emergent field of game composition by exploring how writing in the PC and console editions of Mojang Studio's *Minecraft* reconciles putative antagonisms between nature and technology. I suggest that writing in *Minecraft* models Sidney I. Dobrin and Christian Weisser's concept of 'ecocomposition', which theorises that nature and discourse are co-constituents. Players' discourse consists of the rocks, trees, dirt, water and biological matter, and this discursive matter provides a set of symbols with which they can write shelter, tools and media. Additionally, I suggest that *Minecraft*'s writing technology entangles writing surface and practice with nature by extending the surface of the 'crafting grid' across its environment. I conclude that while *Minecraft*'s model of ecocomposition encourages writing studies of digital games that show the connections between writing, nature and materiality, much of this early eco-game media is entrenched in Cartesian notions of plant, animal and mineral agency that reduces ecological complexity.

In this paper I argue that Mojang's *Minecraft* (2011), a digital game about crafting and surviving with and within nature, models writing as an ecological practice that resolves putative antagonisms between nature and technology. Working between the new materialists' critical materialism and ecocritical studies of writing, I demonstrate that concepts like materialization and ecocomposition provide us with a much needed vocabulary and critical framework for understanding the writing practices in ecologically oriented digital games like Mojang's *Minecraft*. I suggest that such a project is especially germane given that *Minecraft*'s widely circulated ecocompositional model is one of the first in a rapidly growing genre of digital game media overtly using writing to materialise the relations between humans, nature and technology. Thus, while I argue that the model of writing afforded by *Minecraft* can productively expand our definition of ecocomposition and what ecocompositional practice might resemble, I also point towards needed interventions in the ecompositional model depicted in *Minecraft* and its progeny.

I begin with ecocomposition, which asserts nature and discourse are co-constituted through writers and writing, as a core ideology underpinning the design of nature and technology in *Minecraft*. In *Natural Discourse: Towards Ecocomposition,* Dobrin and Weisser (2002) define ecocomposition as a study that examines the relationships 'between environments (and by that we mean natural, constructed, and even imagined places) and discourse (speaking, writing, and thinking)' (6). In short, Dobrin and Weisser argue that nature and the 'natural' are constructed by discourse and vice versa. And although the new materialists have contested this position (a point which I later return to) by arguing for a return to 'the real' outside of discourse, writing technology in *Minecraft* does closely follow the ecocompositional framework posited by Dobrin and Weisser. Players' discourse consists of the rocks, trees, dirt, water and biological matter, and this discursive matter provides a set of symbols with which they can write shelter, tools and media. Digital games like *Minecraft* thus puts pressure on definitions of ecocomposition that reduce it to a field of study or a form of writing *about* nature – adhering more closely to Dobrin and Weisser's understanding of ecocomposition as a writing practice that intersects nature and technology in significant and critical ways. *Minecraft* (Mojang 2009), *Rust* (Facepunch Studios 2013), *Terraria* (Re-Logic 2011), *Starbound* (Chucklefish Games 2013) and forthcoming digital games such as *PixelJunk Inc* (Q-Games n.d.) and No *Man's Sky* (Hello Games forthcoming) in the survival-crafting genre present players with a model of writing that is itself a part of nature and technology, and player participation in these digital environments exposes how technology, nature and space are intimately entangled through writing.

Unpacking materials: studies of discourse, play and materiality

Before launching into an analysis of *Minecraft*, I provide here a brief overview of several digital game studies that have forged a path for theorizing the intersections of discourse and materiality. Although by no means comprehensive, I provide this brief overview for a few reasons. First, I hope to show that my work with *Minecraft* is located in a larger tradition of digital game studies that understands play as procedural discourse that provides an interface for challenging seemingly opposed categories such as human and animal, technology and nature, utopia and reality. Second, and building from play's capacity to collapse binaries, I offer these studies as a way to square the differences between new materialism's emphasis on material realism and ecocomposition's constructivist orientation. Finally, by bringing together work from ecocomposition and new materialism, I hope to resist the vastly overstated promise and premise of the virtual as a digital space in which the players' body disappears and material concerns magically vanish. Instead, in locating and building on game studies that intersect discourse and materiality, I intend to put further pressure on the popular notion that games and the actions of their players are immaterial.

Pushing against the assumption that digital play is purely immaterial, scholars including Consalvo (2009), Flanagan (2013), and Apperly and Jayemane (2012) have shown how play is an activity with material and physical consequence. Much of this work is tied to the field of rhetoric, where theorists like Burke (1966) have articulated play as an activity that unites discourse and materiality. In the wake of critical animal studies, Burke's work with language, animals and the dramatistic in *Language as Symbolic Action* greatly simplifies and, at worst, anthropomorphises how animals communicate; however, these early studies do provide a tentative step towards theorizing how play connects discourse and materiality. Burke applies a 'dramatistic' framework to the study of animals' behavioural communication to suggest that animal behaviours do constitute a classifying discourse in which activities such as nodding or snorting are emblematic of affirmation or negation. Burke (1966) contrasts his 'dramatistic' framework with Henri Bergson's 'scientist' emphasis to demonstrate that with issues of language, 'Whereas Bergson starts from problems of truth or falsity, we start from problems of action' (421). The dramatistic favours the imperative mood rather than the indicative, which is a shift that game studies of procedural discourse have widely adopted, and Burke (1966) uses the dramatistic to hypothesise that animals can indicate the admonitory through an affirmative action. 'Playful' activities such as nipping are both a way of playing, and a discursive representation of a serious act with material consequence that it is not: a bite. Burke (1966) suggests that prior to the emergence of words, 'the negative could be dramatized only by the presentation of stylised and gestural forms of the positive' (423). Thus, play such as a nip is a way of saying 'no', but also a way of communicating the material consequence of a 'bite'. Sutton-Smith (1997), a scholar of play and rhetoric in both humans and animals, draws on Burke's example to suggest that, 'a playful nip is not only not a bite, it is also *not* not a bite. That is, it is a positive, the sum of two negatives. Which is again to say that the playful nip may not be a bite, but it is indeed what a bite means' (1). Within Burke's dramatistic framework, activities such as play are both material and discursive practice that blend the serious and irreverent in complex ways.

Since Burke, many game scholars have returned to the problem of action to further unpack the relationship play configures between materiality and discourse. In Consalvo's *Cheating* (2009), she examines the imperative mood established by the rhetoric of digital games' paratextual industries such as strategy guides and FAQs for playing games, game magazines and legal documents binding players to a code of 'ethical' play within digital games (21). Consalvo argues that these documents grafted their discourses to players and the spaces they occupy through diagrammed maps and strategies that advised players how to follow specific paths and courses of actions. These documents taught players how to play, providing a behaviouristic script that emphasised the importance of specific objectives and methods to achieve them such as high scores and beating bosses (64). Thus, digital game paratexts simplify the Barthean galaxy of signifiers (2000) which 'has no beginning; it is reversible; we gain access to it by several entrances, none of which can be authoritatively declared to be the main one; the codes it mobilizes extend as far as the eye can reach, they are indeterminable' (*S/Z* 5) by narrowing the multiplicity of entrances and paths to a single way of 'mastering' the game. According to Consalvo, these media that taught players how to play and what to expect from a game (25) encouraged ways of thinking about space and the relationship between players and other characters within those spaces. Paratexts reinforce the system of ethics present within digital games through their rhetoric,

> The narration [of strategy guides], however, also keeps the player in the world of the game, to some extent, through the careful use of language. Drawing from the world of Tom Clancy, enemies are 'hostiles' and 'tangos.' The player is encouraged to 'eliminate' such terrorists, who of course don't really 'greet you' so much as attempt to shoot you on sight. (61)

The attempt at consistency between the paratext's discourse, the player's actions, and the narrative/procedural discourse within the game bears a striking resemblance to the ecocompositional practice described by Dobrin and Weisser (2002) as a discursive construction of natural, constructed and even imagined places (6). Consalvo's study (2009) shows how digital place, environment, space, site, location affect written discourse as well as the player's procedural discourse within those spaces, showing the role that writing plays in constructing digital environments and vice versa.

While many digital game studies look at how games and their paratexts inform players' relationship to digital and non-digital space/individuals, Apperly and Jayemane (2012) have built on this work to perform a specifically new materialist study of digital games and how these media impact the environment. Apperly and Jayemane remind readers that players and games do not vanish into thin air once they begin playing, but that

> Digital games are objects that exist in the world; however much their digital virtuality is celebrated they are enacted and produced in strikingly visceral – ontologically virtual – ways. The noise that a PS3 game makes when it is pushed into the blu-ray drive, the cables over which Xbox LIVE arcade games are downloaded, and the wars and environmental depletion taking place to produce the coltan-based batteries in your wireless devices … all suggest a materiality to gaming that links digital games to the world and demands that they are also understood as objects in the world (15).

Apperly and Jayemane observe that as players play within the game's galaxy of signifiers, digital games produce physical and digital e-waste, and this issue has been confronted by game designers like Nifflas whose *Knytt Stories* (n.d.) is specifically designed to consume as little processing power as possible. Players also confront materiality during play through prompts requiring them to spend real money for 'virtual' goods, through game controllers that are designed for bodies capable of certain actions, through intense affective feelings for supposedly 'virtual' characters, through motion controls and cameras respond to bodily motion and gestures and through the loss of characters in which they have invested significant amounts of time. In short, play has never been entirely separate from discursive and material consequences, and the emergence of digital technology has coupled the two even more closely.

Apperly and Jayemane's material analysis of digital games demonstrates that the game's textuality extends well beyond the screen and includes everything from the faint sound of cooling fans, to the labour exploitation and wars sparked by the console's conflict minerals. Yet while Apperly and Jayemane extend digital game text to include a console's external materiality, internally there occurs a complex play between material hardware and data that produces the galaxy of signifiers that players directly interact with throughout their play. This internal interaction between hardware and data occurs on at least three levels. First, code defines what an in-game object *does*, and when digital objects *do* something it requires more physical memory resulting in increased energy consumption. Next, the images and 3D/2D models that we look at when playing provide players with a visual referent for the clusters of code chugging in the backend, and these models require physical space on computer hard drives and significant CPU power to render. Third, digital games involve a material literacy between machine and player: the machine responds to physical input from players by reading code, and players actions are a response to state changes within the machine as visualised on screen (or via other haptic output). This trifecta, although in no way complete, demonstrates that digital games' materiality are foundational to their discursivity. Thus, whenever we discuss play we are also talking (albeit indirectly) about digital games' materiality and discursivity, meaning that ecocomposition and new materialist work are extremely relevant to the study of digital games.

Minecraft and ecocomposition: writing nature and nature writing

Minecraft, officially released in 2011, is a game in which players create and build tools, objects and structures out of blocks. Much like a set of nature themed Legos, each block is a different type of matter such as 'dirt', 'water', 'wood', 'lava', that are stuck together to create the in-game environment (Figure 1). Moreover, these blocks compose the 3D, randomly generated game world, and they can aid (or hinder) the player with activities such as exploring, gathering resources, building, crafting and combat. Although there are many nuanced ways these blocks interact with players and vice versa, players have two primary actions they can perform with them: breaking blocks or placing them back into

Figure 1. A highlighted block of sand depicting how the environment consists of discrete, individual blocks placed side-by-side. Source: Minecraft ®/TM & © 2009–2013 Mojang / Notch.

the world. When the blocks are broken, they release a tiny cubic 'symbol' that represents the properties and dimensions of the block. Players break blocks in the world to collect these symbols, and they can either 'craft' tools and objects with the symbols, or 'place' them back in the world to build shelter or other media (Figure 2).

Unlike other digital games, in which nature and ecologies are situated at the game's margins as scenic backdrops for grand conflicts between players and other humans/technology, digital games within the survival crafting genre model a nature and ecosystem within which players are deeply embedded. To survive, players must practice sustainable (and, as I later discuss, oftentimes not so sustainable) practices within the environment.

Figure 2. Symbols that players can collect after breaking blocks in the environment. Source: Minecraft ®/TM & © 2009–2013 Mojang / Notch.

Nature is not a passive agent, however, and creatures and the environment will act in ways that subvert and challenge the player's attempts to harvest resources and build shelter. Lava flows, flooding, creatures and other players on the same server can quickly kill and destroy anything that players have built. Additionally, the player's character has a hunger meter that depletes as players perform activities (running and jumping deplete this meter even faster), which means that even long walks or cave spelunking can prove deadly for the unprepared player. While many of these actions are player-oriented, *Minecraft* and other survival craft games include non-playable characters (NPCs) such as 'Creepers', 'Endermen' and 'Villagers', that will move and destroy blocks and carry items, leading to changes in the environment that occur outside of the player's control. *Minecraft* models an agential nature that, unlike many digital games, challenges players to develop practices that enable them to survive within.

Survival, however, does not come down to a zero sum game of having the best tools. The ability to exist as a member of this ecosystem greatly depends on how players intersect nature and technology through what I will argue are ecocompositional practices. The ability to write with and within the environment has been harnessed by players to transform *Minecraft* into an experimental space for writing and circulating several types of media including *Machinima*, complex environmental mods, to scale replicas of iconic architecture, micro-computers and even new games. Whereas most media production software de-materialises media production's effects on the environment, *Minecraft*'s player community has re-coupled media production with ecological concerns by *writing* media with mineral resources, ecosystems, creatures, food and other matter. So while survival depends on players' ability to live off of the elements in the game's biosphere, which Dobrin and Weisser (2002) define as 'consisting of the earth and its atmosphere, which supports our physical existence' (13), players have also developed a paratextual semiosphere 'consisting of discourse, which shapes our existence and allows us to make sense of it' (13) that they are encouraged to contribute to and circulate with Youtube videos, FAQs and servers that host media produced from this digital environment.

The building blocks of ecocomposition in *Minecraft* are the blocks that represent the game environment. One hundred fifty three block types exist as of version 1.7.4, and each block symbolises a different type of material such as dirt, water, stone, wood, ores, sand, wool, etc. (Figure 3). *Minecraft* includes sixty-one unique biomes, which are distinct areas differentiated by specific materials and flora that only appear in these areas. The 'Roofed Forest' biome, for example, is composed of dark oak blocks, leaf blocks, mushroom blocks, dirt blocks, sand blocks and water blocks; the leaf blocks that grow from the dark oak trees are spaced tightly together to create a dense canopy that makes the ground dark even during daytime (Figure 4). The 'Mesa (Bryce)' biome, on the other hand, consists of sand blocks, cactus, hardened clay blocks, six types of stained clay blocks and dead bushes (Figure 5). The landscape here mimics the real Bryce Canyon with a low desert-like ground region and tall, spire-shaped columns of hardened clay blocks. This patchwork structure is repeated within each biome as its blocks are arranged into structures that resemble trees, sandy beaches or underground caves.

The underlying grid that structures these blocks into a legible environment is significant to how *Minecraft* intersects writing technology with nature. Blocks in *Minecraft* are arranged in a fixed grid pattern, which means players can only place blocks and objects on fixed locations relative to the grid. Blocks consist of six sides, and players can only place additional blocks or objects parallel to one of these sides. In other words, players can only place blocks to cover the entire area of a block's side – they cannot cover

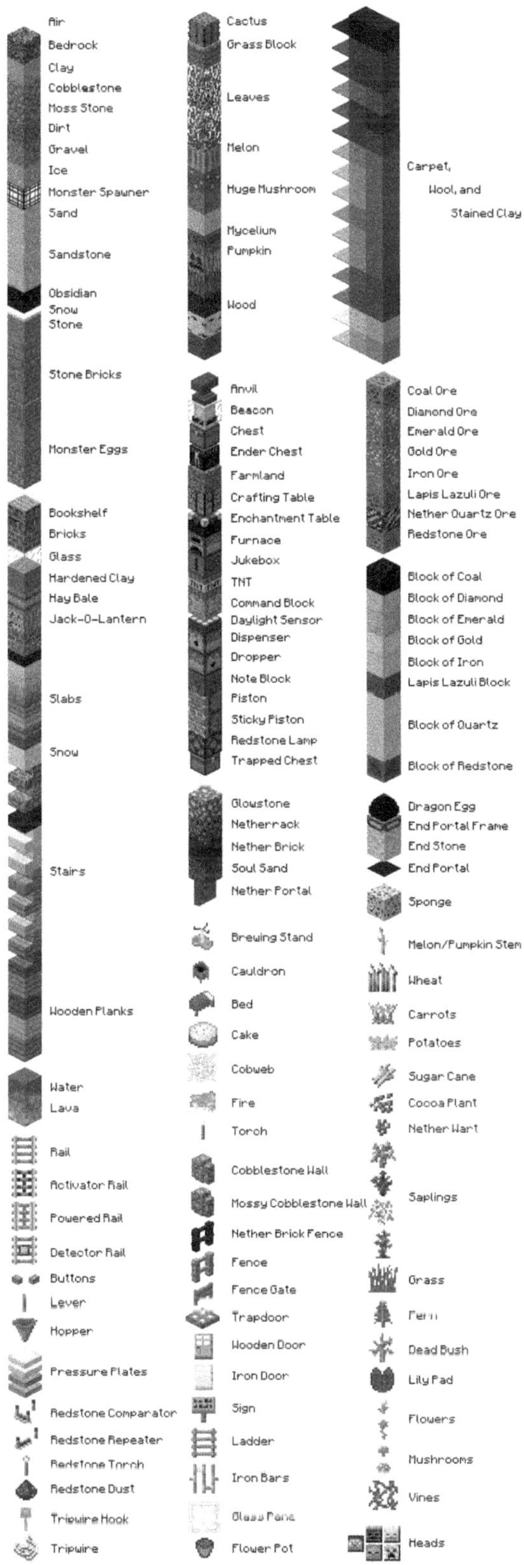

Figure 3. Hunter712's (2013) list of block types. Source: Minecraft ®/TM & © 2009–2013 Mojang / Notch.

Figure 4. Roofed Forest Biome (captured by Minecraft forum user KowaretaTamashi (2013)). Source: Minecraft ®/TM & © 2009–2013 Mojang / Notch.

Figure 5. Mesa (Bryce) Biome (captured by oCrapaCreeper (2013)). Source: Minecraft ®/TM & © 2009–2013 Mojang / Notch.

a quarter or half of a side. Much like Legos, objects and blocks 'snap to' the grid and cannot be placed freely.

Whereas in-game writing technologies such as the typewriter in *Resident Evil* (1996) or the cell phone in *Grand Theft Auto 4* (2008) are procedurally and visually distinct from the game's environment, the environment's grid structure in *Minecraft* is mirrored in the 'crafting interface', which entangles environment and writing technology (Figure 6). Although formally called the 'crafting interface', this 2 × 2 and 3 × 3 grid is a simplified writing pad for symbolizing tools, objects and items with resources from the environment. To write tools and items in *Minecraft*, players must first collect the cubic symbols that appear after breaking a block (Figure 2). Upon collecting these symbols, they move into the player's inventory (Figure 7). The inventory's 3 × 9 storage grid houses the cubic symbols as 2D squares with a counter in the right-hand corner of the image that denotes the number of cubes stacked into a single inventory slot. Whereas this grid is only visible

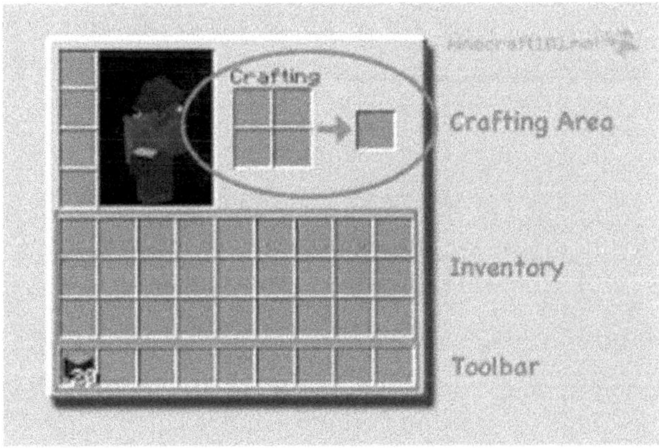

Figure 6. Player inventory and crafting interface. Source: Minecraft ®/TM & © 2009–2013 Mojang / Notch.

in the environment when the player highlights a block with her character's hand, leading to the (mis)perception of a unified nature, the inventory grid chunks these blocks into discrete compartments independent of the other block types.

Above the player's inventory storage space is a 2 × 2 'crafting grid' that can produce simple objects such as wooden planks or a crafting table. Each block contains a symbolic value that is extracted by crafting, and by experimenting with different patterns and blocks in the crafting interface, players discover that blocks function as both environment and symbols with which they can write tools and items. Wood blocks, for example, moved from the inventory into any region of the crafting grid causes 'wood planks' to appear in the square to the right of the arrow (Figure 8). The player can then move the wood planks into her inventory. Removing the wood planks from the slot, however, will consume the wood blocks

Figure 7. Materials in the player's inventory (captured by Shine53573 (2012)). Source: Minecraft ®/ TM & © 2009–2013 Mojang / Notch.

Figure 8. Crafting wood planks. Source: Minecraft ®/TM & © 2009–2013 Mojang / Notch.

in the crafting grid, and crafted items cannot be recycled for materials. Moreover, whenever a player crafts an item or material in *Minecraft*, the materials that the player used to craft that item or material will be consumed in the process and permanently erased from the world.

Once players unlock the expanded 3 × 3 crafting grid by building a 'crafting table', they can then arrange harvested resources in a pictographic pattern that resembles those items such as a 'T' or a ' = ' shape to produce tools and objects such as a pickaxe, torch or a bed. Unlike the arrangement of squares in the 2 × 2 grid, the arrangement of squares in the crafting table's 3 × 3 grid is non-trivial and must mimic the structure of the object that the player wants to produce (Figure 9). These abstract 'signs' signify an object or tool to the program, which it produces for the player in the right-hand box only if the player has used the correct materials and organised them in the correct pattern. For example, players cannot make a bed by placing three iron ore in a horizontal row and three wood in a horizontal row below; instead, players must use three wool and three wood arranged in a pattern that both player and game recognise as 'bed' (Figure 10).

Most significant, however, is the way that the structure of the in-game crafting grid actually parallels the fixed grid that fits together the blocks that compose the game's environment. These blocks are organised into a coherent environment through the use of a fixed grid that fits blocks together such that they resemble a beach or a tree. The structure of the 'crafting area' follows the same grid arrangement as the game environment, intimately entangling writing technology with the environment without eliminating the differences of either. The crafting grid's structure is the trace of the external environment that haunts the moment of writing and resource consumption in *Minecraft*, revealing that

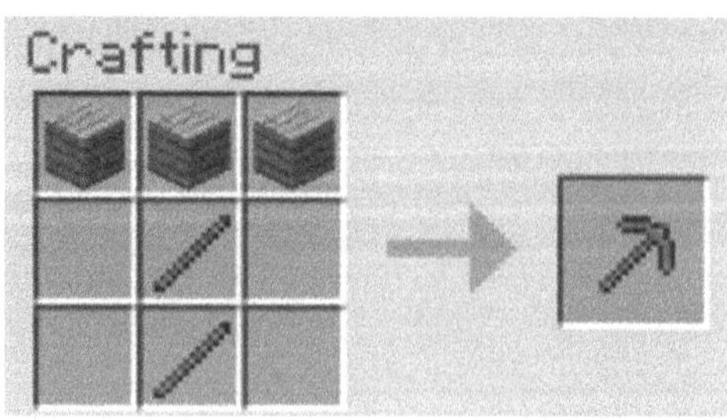

Figure 9. Pattern for crafting a pickaxe. Source: Minecraft ®/TM & © 2009–2013 Mojang / Notch.

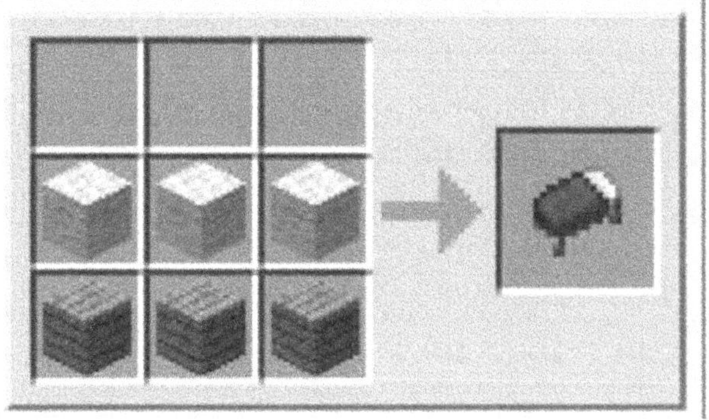

Figure 10. Pattern for crafting a bed. Source: Minecraft ®/TM & © 2009–2013 Mojang / Notch.

the entire space of *Minecraft* is a writing technology. Thus, the environments of digital games like *Minecraft* enable us to experience writing as an always already ecocompositional practice, and *Minecraft* gives us a way to talk about the material and terminal qualities of writing. Writing, as theorised in *Minecraft*, is a practice of patterning nonrenewable resources, and this nonrenewable practice argues that writing has material limits.

While the crafting grid is used to write tools and objects within the game's biosphere, its part to whole relationship with the larger game environment shows us that the environment itself is a writing technology that players have used to build a semiosphere consisting of larger media such as castles, 16-bit computers and death star replicas. As players take on larger projects, they must find and extract more resources from the environment to build that media, which leaves sizable scars on the game's terrain. Some of these traces are demonstrated by fan videos, which show vast destruction and transformation of the environment (Figures 11 and 12), showing how writing materials and writing itself leaves a permanent mark on the world.

Figure 11. How a 16-bit advanced logic unit affects the environment in *Minecraft*. Source: Minecraft ®/TM & © 2009–2013 Mojang / Notch.

Figure 12. 1:1 scale model of the USS Enterprise showing entire landscape stripped away. Source: Minecraft ®/TM & © 2009–2013 Mojang / Notch.

Minecraft's semiosphere challenges non-ecological models of writing presented in media design platforms such as MS Word or Photoshop that de-materialise the producer's entanglement with and dependency on the environment, where the spatial and environmental dimensions of media production are abstracted to empty, isolated canvas on which media can be printed and erased with a single keystroke. Many design programs allow operators to produce media simply by tapping keys or dragging the pointer around the screen, which promotes an image of composition as divorced from nature. Emerson (2014), in *Reading Writing Interfaces*, argues that the standard design of digital writing interfaces such as word processors, video-editing suites and computer-aided design platforms is intended to conceal how media production and writing affect and use nature while simultaneously branding themselves as 'natural'. What 'natural' means in most interface design is 'a belief in the value of an interface that recedes from view, ideally to the point of invisibility, which now also implies inaccessibility. We need not know how it works, or how it works on us rather than us on it' (6). van Dam (1997) displays this sentiment in his design manifesto, 'Post-WIMP User Interfaces:'

> However user-friendly, an interface is still an intermediary between the user's intent and execution of that intent. As such, it should be considered at best a necessary evil because no matter how fluid, it still imposes a layer of cognitive processing between the user and the computer's execution of the user's intent. The ideal interface is no interface – 'I think, therefore the computer gives me what I thought about (and what I should think about)'. (64)

Unsurprisingly, van Dam later describes this 'interface that is no interface at all' with the phrase 'natural', when it is in fact, and of course, manufactured. What *Minecraft* exposes, however, is that abstracting or removing the crafting interface would actually eliminate a significant point for configuring the relationship between humans, writing and nature. Van Dam's insistence on identifying and re-producing a 'natural' experience of the world with post-WIMP interface in fact stresses the differences between humans, technology and nature, presenting humans and nature apart from technology and not a part of technology. As Emerson argues, post-WIMP operators never confront their tools or their tools' effects on nature in the way that *Minecraft* presents to its players.

In terms of digital writing technology, signifying materials, such as images or alphabetic text, are not entangled with the environment. Instead, they seem to appear *ex nihilo* from an operator's fingers. *Minecraft*, instead, defamiliarises the typical digital writing environment with an environment from which writers must extract materials to signify with. Images, structures and alphabetic text can only be made with materials collected from the environment. Instead of simply clicking on a dropdown menu to access a colour and style of font, players must seek out these materials by reading the caves that they have carved out, perhaps recalling that they have depleted gold ore and red stone from the caves in the south and now must head north. Biomes such as deserts and jungles present opportunities for different resources, which encourages players to think of the different possibilities for media production that each environment affords. Players who have devastated the environment to produce a 16-bit ALU have eliminated the possible ways they can interact with the environment. In Figure 12, the player can no longer sit under the shade of trees, and the darkness of cave systems cannot draw her into the depths of the world, but building a model of this machine in AutoCAD would have none of those consequences. The environment in *Minecraft* inscribes itself on the player as much as the player inscribes herself upon it, and this ecocompositional model is unique to digital games like *Minecraft*.

Minecraft and new materialism: changing the environment

I conclude this paper with a brief new materialist critique of ecocomposition in *Minecraft* to (hopefully) demonstrate the complexity of digital natures and encourage further discussion about ecocomposition in digital games. I believe *Minecraft* and its progeny have taken great strides from digital nature as a scenic, non-vital backdrop to a significant actor that players must work with and within, but there remains much work to do because despite offering a productive challenge to think of ecocomposition as practice and challenge how most writing interfaces obscure the relationship between writing and nature, *Minecraft*'s model of ecocomposition, like many models that quantify nature and ecology, does slip into a reductive Cartesianist treatment of matter that greatly reduces ecological complexity. In Coole and Frost's (2010) introduction to *New Materialisms: Ontology, Agency, and Politics*, they suggest that 'Many of our ideas about materiality in fact remain indebted to Descartes (2008), who defined matter in the seventeenth century as corporeal substance constituted of length, breadth, and thickness; as extended, uniform, and inert' (Coole and Frost 7), and that this perception of matter has encouraged a metrification of nature as quantifiable and measurable. 'According to this model, material objects are identifiably discrete' – notably non-ecological materials independent of any relation with other matter – 'they move only upon an encounter with an external force or agent, and they do so according to a linear logic of cause and effect' (7). Much like the blocks in *Minecraft*, Cartesian matter is controllable and predictable, bound to laws of Euclidian geometry and Newtonian physics, bequeathing human subjects a sense of mastery over the land rather than truly positioning them as a member of it.

Moreover, while I have attempted to argue that digital and non-digital play always intersects discourse and materiality, Coole and Frost are concerned with these discursive orientations to objects and nature given their sense that it promotes 'an allergy to "the real" that is characteristic of its more linguistic or discursive forms ... has had the consequence of dissuading critical inquirers from the more empirical kinds of investigation that material processes and structures require' (6). Although I certainly align with ecocomposition's emphasis on the intersection of discourse and materiality to theorise how players work with and within nature to produce discourse (Dobrin and Weisser 2002, 9–10), I agree with Coole and Frost that we must be critical of how the digital 'galaxy of signifiers' in *Minecraft* promotes Cartesian ideologies about nature and matter that we should challenge. Additionally, *Minecraft*'s global 'naturalization' of discourse erases Dobrin and Weisser's (2002) key point that discourses are deeply connected to the circumstances and locations from which they emerge. Although players can farm different types of sand, for example, that difference is purely visual and in no way is reflected in differences in how the material behaves or what can be produced with it.

In *Vibrant Matter*, Bennett (2010) proposes that new materialist and ecocritical work should embody a 'vitality of matter', which demonstrates 'the capacity of things – edibles, commodities, storms, metals – not only to impede or block the will and designs of humans but also to act as quasi agents or forces with trajectories, propensities, or tendencies of their own' (viii). Instead of asking what we can do *to* matter, a question that *Minecraft* explicitly puts to its writers, ecocritical vitality rethinks matter's mattering in terms of what its forces and actions can do to surrounding material, including physical, political and sociocultural events. Composing with vital material means working with material that has 'efficacy, can *do* things, has sufficient coherence to make a difference, produce effects, alter the course of events' (viii). This definition of ecocomposition challenges images of 'dead' composition with scenes of vibrant, vital matter that collaborates to create the possibility for more ecologically and materially sustainable modes of production and consumption.

Unlike the vital material that Bennett calls for, *Minecraft*'s writing interface instantly produces items from resources, which abstracts the actual physical labour and waste involved when working materials into a tool. By concealing the player's hands and body with the crafting grid overlay, this interface also suggests that the interface alone does the work of destroying natural materials thereby lessening feelings of responsibility for how players use the environment. Breaking and making tools produces no waste, which, while encouraging playfulness, also works to minimise an ethics of responsibility for the depletion of environmental resources. Designers, however, have not ignored *Minecraft*'s issues, and the recent multitude of survival crafting games such as *Rust, Starbound*, *Terraria*, and the forthcoming *No Man's Sky* and *PixelJunk Inc.* have begun to offer even more nuanced understandings of the relationship between nature and technology through ecocomposition.

References

Apperly, T., and D. Jayemane. 2012. "Game Studies Material Turn." *Westminster Papers* 9 (1): 4–25. https://www.westminster.ac.uk/__data/assets/pdf_file/0009/182448/002.-Game-Studies-Material-Turn-Thomas-H.-Apperley-and-Darshana-Jayemane.pdf.

Barthes, R. 2000. *S/Z: An Essay*. New York: Hill and Wang.
Bennett, J. 2010. *Vibrant Matter: A Political Ecology of Things*. Durham, NC: Duke University Press.
Burke, K. 1966. *Language as Symbolic Action: Essays on Life, Literature, and Method*. Berkeley: University of California Press.
Capcom. 1996. *Resident Evil*. PC, Mac, and Console game.
Chucklefish Games. 2013. *Starbound*. PC and Mac game.
Consalvo, M. 2009. *Cheating: Gaining Advantage in Videogames*. Cambridge, MA: The MIT Press.
Coole, D., and S. Frost. 2010. *New Materialisms: Ontology, Agency, and Politics*. Durham, NC: Duke University Press.
Descartes, R. 2008. *A Discourse on Method*. Project Gutenburg. http://www.gutenberg.org/files/59/59-h/59-h.htm
Dobrin, S. I., and C. Weisser. 2002. *Natural Discourse: Towards Ecocomposition*. New York: State University of New York Press.
Emerson, L. 2014. *Reading Writing Interfaces: From the Digital to the Bookbound*. Minneapolis: University of Minnesota Press.
Facepunch Studios. 2013. *Rust*. PC and Mac game.
Flanagan, M. 2013. *Critical Play: Radical Game Design*. Cambridge, MA: The MIT Press.
Hello Games. forthcoming. *No Man's Sky*. Console and PC game.
Hunter712. 2013. "List of Block Types." Image Table. *Minecraft Wiki*, December 17. http://minecraft.gamepedia.com/index.php?title=File:Template1.png&offset=20110106235903
KowaretaTamashi. 2013. "Roofed Forest Biome." Screenshot. *Minecraft Forum*, September 5. http://www.minecraftforum.net/forums/minecraft-discussion/recent-updates-and-snapshots/future-updates/384775-found-something-very-mysterious-roofed-forest-a
Mojang. 2009. "Inventory Screen." Screenshot. *Minecraft101.net*. http://www.minecraft101.net/reference/images/Inventory-screen_thb.jpg
Mojang. 2011. *Minecraft*. Stockholm: Mojang, PC and Mac game.
Nifflas. n.d. *Knytt Stories*. PC and Mac game.
oCrapaCreeper. 2013. "Mesa (Bryce) Biome." Screenshot. *Minecraft Forum*, October 2. http://www.minecraftforum.net/forums/minecraft-discussion/recent-updates-and-snapshots/381982-mesa-bryce
Q-Games. n.d. *PixelJunk Inc*. Console and PC game.
Re-Logic. 2011. *Terraria*. PC, Mac, and Console game.
Rockstar North. 2008. *Grand Theft Auto 4*. Console and PC game.
Shine53573. 2012. "Inventory Image." Screenshot. *Minecraft Wiki*, May 25. http://minecraft.gamepedia.com/File:2012-01-23_16.42.57.png
Sutton-Smith, B. 1997. *The Ambiguity of Play*. Cambridge, MA: Harvard University.
Van Dam, A. 1997. "Post-WIMP User Interfaces." *Communications of the ACM* 40 (2): 63–67. doi:10.1145/253671.253708.

Rub trees, webcams, and GIS: the wired wilderness of Leanne Allison and Jeremy Mendes' *Bear 71*

Sarah Jaquette Ray

Environmental Studies, Humboldt State University, Arcata, USA

This article analyses Leanne Allison and Jeremy Mendes' 2012 interactive documentary, *Bear 71*. Supported by the National Film Board of Canada, Allison and Mendes made use of thousands of hours of National Park webcams footage of wildlife to inform audiences that, in Banff, 'bears and humans live closer to each other than any place in the world'. According to Sundance, this multimedia work is a 'poignant interactive documentary about a bear in the Canadian Rockies [that] illuminates the way humans engage with wildlife in the age of networks, satellites, and digital surveillance'. Although *Bear 71* has received no scholarly attention, it explicitly engages concerns that are central to animal studies, interdisciplinary environmental humanities, critical geography, and eco-media studies. The navigable digital environment that the text simulates and produces, represented through GIS technology, the videogame-like interactivity of that space with your own 'avatar', and the webcam access to intimate, everyday animal life, all raise questions about how new media produces knowledge about nature, the role of mapping tools such as Geographic Information Systems and scientific technologies like radio telemetry and crittercams in shaping human–animal relations, and the viability of an environmental ethic that bemoans the loss of nature 'out there'. Drawing on scholarship about crittercams and wildlife, such as Benson's *Wired Wilderness* and Donna Haraway's *Crittercam*, and critical animal geography – especially what Sarah Whatmore calls 'hybrid geography' – I argue that the film's use and representation of interactive digital environments reveal the ways that humans and nature are entangled. The film forces us to imagine that we always already live in 'material-semiotic' worlds and asks us to consider how 'the wild and the domestic get swept up in the volatile eddies and flows of socio-technical networks that bring people, living organisms and machines together in new and particular ways' (Whatmore, 344). The implications for understanding *Bear 71* as articulating a hybrid geography demonstrate the contribution of critical eco-media studies to environmental thinking: space is not material, fixed, or objectively knowable; the human and the wild constitute and attenuate each other; and technology is central to – as opposed to corrupting or only mediating – our ability to perceive how we already participate with and within nature.

'Sometimes it's hard to tell where the wired world begins and the wild one ends'.

–*Bear 71*

'If wildness can stop being (just) out there and start being (also) in here, if it can start being as humane as it is natural, then perhaps we can get on with the unending task of struggling to live rightly in the world' (90).

–William Cronon, 'The Trouble with Wilderness'

Figure 1. Screen shot of *Bear 71* advertisement.
Source: Bear 71 ©2012 National Film Board of Canada.

Leanne Allison and Jeremy Mendes' 2012 open-access interactive web documentary, *Bear 71*, exemplifies the current tensions between technology and conceptions of the wild that concern many environmentalists. The 'film', which is part film, part interactive video game, part website (hence, 'web documentary' or 'webdoc'), tells the first-person story of a radio-collared grizzly bear in Banff National Park in Canada (see Figure 1). It makes artistic use of webcam footage provided by the Park of the lives of the creatures that inhabit the wilderness landscape. 'Users' (as opposed to 'viewers') of the webdoc can navigate a map of Banff to learn more about its animal species and landscape features, and even follow bear 71 herself. The webdoc uses a variety of technologies, particularly the worldwide web, webcam photography, surveillance networks, and multiple multimedia platforms accessible on personal computers, to convey a story about a bear using alternative media forms. The form also makes use of digitalised representations of nature, particularly wildlife webcam footage and radio-tracking and Geographic Information Systems (GIS) mapping (see Figure 2). The webdoc thus seems to celebrate networks, technological convenience, observational and mapping technologies, and user-friendliness. It appeals to audiences who are technologically savvy, connected by multiple devices and networks to people and issues far and near, and who benefit from information technologies connected by globalisation.

But the voiceover, told from the bear's perspective and written by *100-Mile Diet* co-author J.B. MacKinnon (MacKinnon and Smith 2007), articulates a much more ambivalent view of all kinds of human encroachment, including the technologies associated with monitoring and portraying wild animals and with globalisation more broadly. Using media platforms that 'appeal to the young urbanite' (Lederman 2012), as one reviewer put it, the film tells the story of the end of wilderness through the biography of bear 71, who ultimately meets her fate on train tracks. The bear's first-person narrative is elegiac about wildness lost and posits a wilderness home encroached upon by

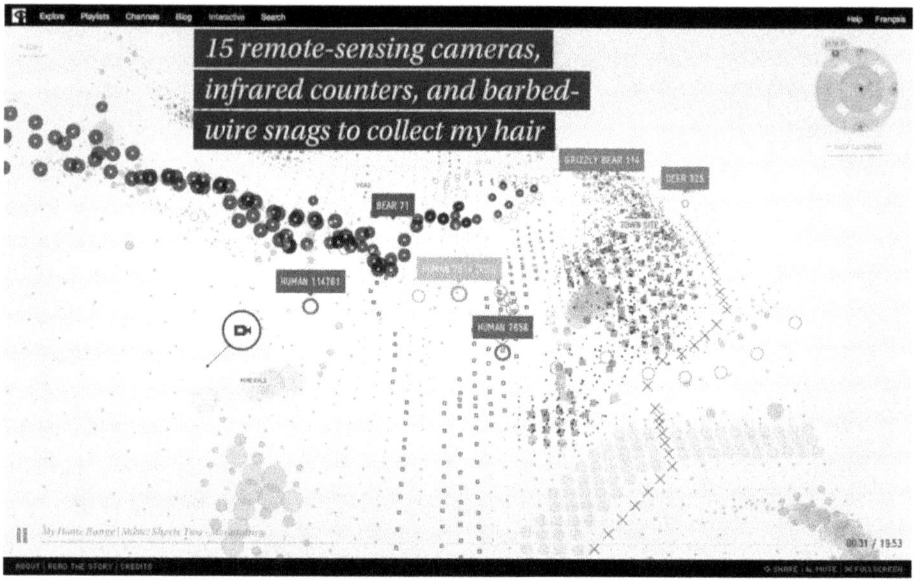

Figure 2. Bear 71 showcases technologies of both wildlife observation and new media.
Source: Bear 71

highways, tourism, globalised transport systems, development, and even the very mapping and surveillance technologies that the webdoc uses to tell bear 71's story. Explicitly designed to diverge from conventional genres of nature programming (such as Discovery Channel's 2001 hit series *Blue Planet* [2001], narrated by David Attenborough and Sigourney Weaver), the film paradoxically celebrates both technology and a predigitised 'wild' world. But unlike *Blue Planet*, which was so successful in part because it used and showcased cutting-edge forms of videography, *Bear 71's* form challenges the genre of nature program and centres on questions about the relationship between technology and wilderness.

In interviews, Leanne Allison, too, worries about the loss of wilderness around Banff National Park, the habitat of bear 71 (2012). She portrays a sense of sadness about the loss of the wild near Canmore, where she lives. She and her husband, Karsten Heuer, are avid conservationists, and Allison's first film, *Being Caribou* (2005), attests to her love of nature and her desire to inspire empathy for charismatic megafauna through her filmmaking. As a result, *Bear 71's* highly technological, postmodern, and interactive form, as well as its portrayal of the highly wired wilderness of Banff, would seem to contradict both the narrative of the text and Allison's own notions of unsullied wilderness. In this article, I explore this tension and argue that ultimately, and perhaps counter-intuitively, the webdoc's form, which deliberately blurs boundaries between wired and wild, offers a more constructive environmental sensibility than the narrative's nostalgic position and Allison's own creative intentions can achieve. Indeed, I believe that, despite (or because of) its paradoxes, *Bear 71* helps us better understand the role of surveillance and information technologies in mediating the relationships between humans and nature in an ever more globalised world.

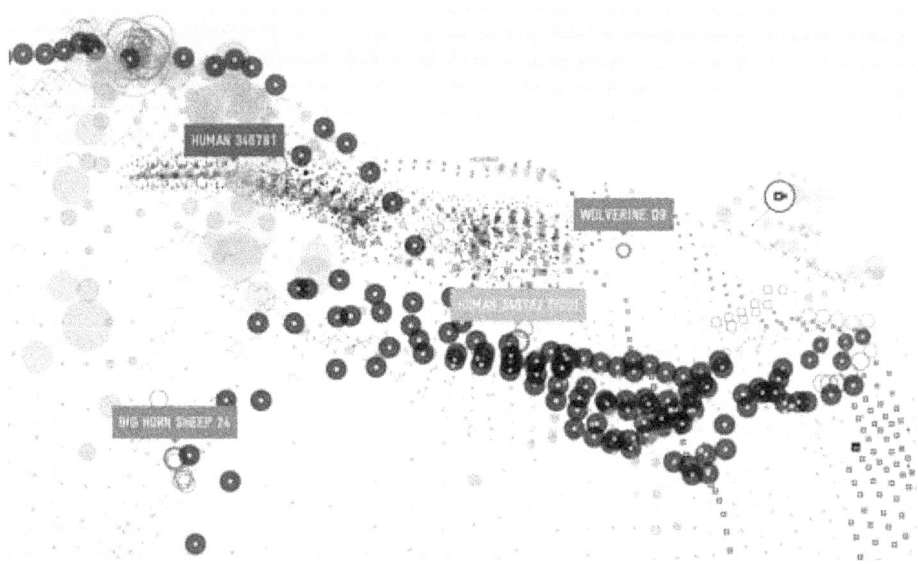

Figure 3. Banff blurs boundaries between wired and wild. Upper left red circles represent a rub tree. Black circles represent the highway and trains moving. Cameras are located throughout the Park. Human viewers of the webdoc are virtually embedded in the landscape.
Source: Bear 71 ©2012 National Film Board of Canada. All rights reserved.

One insight the webdoc provides is that Banff, the epitome of Canadian wilderness, is a profoundly networked, 'wired', studied, documented, and travelled space (see Figure 3). Just as environmental historians like William Cronon (1995), Richard White (1996), Carolyn Merchant (2010), Karl Jacoby (2003), and Mark Spence (2000) have demonstrated, there is no such thing as a truly pristine, untouched wilderness, in Banff or elsewhere. Bear 71's story dramatises this reality from an animal's perspective. With support from the National Film Board of Canada, Allison and Mendes sifted through thousands of hours of Banff webcam footage of wildlife to create a work that informs audiences that, in that wilderness space, 'bears and humans live closer to each other than any place in the world'. According to Sundance, where the film won the prestigious New Frontiers Award, this multimedia work is a 'poignant interactive documentary about a bear in the Canadian Rockies [that] illuminates the way humans engage with wildlife in the age of networks, satellites, and digital surveillance' (Sundance, 2011). The 'text' produces a navigable digital environment using GIS technology to create a videogame-like interactivity with a fictionalised landscape of Banff. A cast of animal characters inhabits this space, and users can click on webcam images of different animals to learn more about them.

In addition, the site automatically activates its user's own computer webcam and places an image of the user within the landscape itself. Thus, concurrent users can see each other and interact simultaneously in the virtual landscape. The viewer thus becomes both subject and object of observation (see Figure 4). This feature feels particularly intimate and allows the user to draw parallels between the conservationist's webcam surveillance of animals and the disciplinary surveillance of human activity that characterises postmodern social life. It also implicates the user in the narrative; as I watch, I begin to imagine that I personally have a material stake in this landscape, even if I've

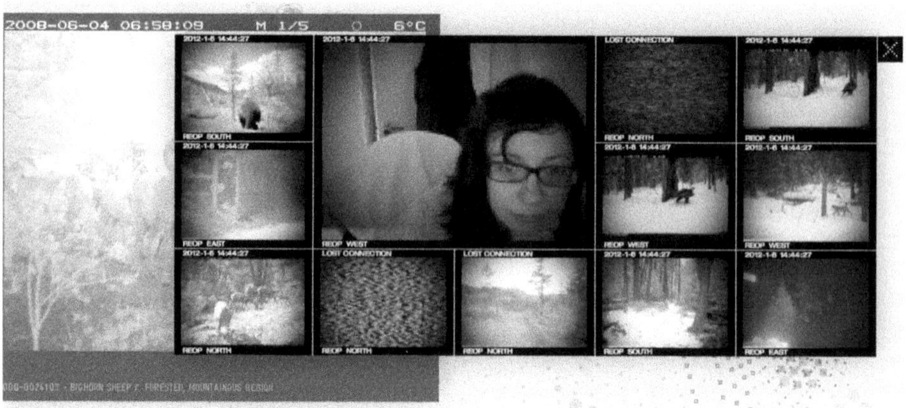

Figure 4. Viewer's own webcam renders viewer simultaneously subject and object of observation. Source: Bear 71 ©2012 National Film Board of Canada. All rights reserved.

never visited it. This webdoc thus raises important questions about how new media produces knowledge about nature and about the role of mapping tools, such as GIS, and of scientific technologies, such as crittercams, in shaping human–animal relations. It exemplifies the role of new media in helping individuals understand their relationships to nature in a globalised, digitised world, where our actions reverberate across networks and seem to transcend geography. Finally, even as the webdoc reflects rather than reconciles the tension between pro-wilderness and post-wilderness poles, it succeeds in arguing that some 'third way' may be needed.

Indeed, in both its multimedia form and its storyline content, the webdoc emerges from and interrogates the binary between the wired world, on the one hand, and the wild world, on the other. On the surface, the webdoc seems to reinforce dominant notions of the wild in a variety of ways. It fits Sarah Whatmore's (2002) description of Western environmentalism. As she writes, 'the co-incidence between "wild" plants and animals (species) and the "wild" spaces they inhabit (inhabits) pervades Western environmental sensibilities' (343). That is, according to dominant cultural convention, wild animals, like the grizzly bear, belong in wild spaces, like Banff. Deconstructing the metonymy between a particular species and a particular landscape aesthetic is crucial to understanding *Bear 71*, as the webdoc's treatment of the bear and its representation of Banff as a geography of wilderness are analogous. Like the grizzly bear, a symbol of the wild, the wilderness itself 'stands as a transcendent sign and site of the radical otherness of a nature without a past; an immaculate space defiled by any taint of human presence' (Whatmore 2007, 13). Leanne Allison herself, whose views I was privileged to hear in two presentations on the film and in two visits to classes in which I taught the film at the University of Alaska Southeast, clearly holds these sensibilities about both wild animals and wild spaces. As the webdoc's opening text tells us, 'there are few ways for a grizzly bear to die. Or at least that's how it was in the wild'. The use of the past tense 'was' conveys nostalgia for the way nature 'used to be' before human penetration of the wilderness multiplied the ways that bears could die. Throughout the narrative, the wild and the wired are contrasted in these poignant ways, suggesting that the filmmakers are making the normative claim that this binary ought to be upheld.

For example, a 20-minute first-person voice-over narrative of bear 71's story and a series of webcam scenes from her life play alongside, and at times disrupt, the user's control of the site, insisting on telling bear 71's story. From bear 71's initial capture at age 3, tagging, and release into Banff, the narrator describes how the bear must learn how to navigate this landscape, which, it becomes clear, is no longer 'natural' or wild for a bear. Thus, taught to survive by '*not* doing what comes naturally' (*Bear 71*), bear 71 has to unlearn her instincts and adapt to a new kind of wilderness that threatens her existence. Becoming adapted to humans, finding spilled grain on railroad tracks, and crossing highway overpasses to get outside the park are some of the ways that wildlife navigate the 'unnatural' spaces of national parks. We learn that bear 71 has cubs, interacts a few times with humans, and more than once narrowly escapes death.

The memoir-style, anthropomorphising narrative makes bear 71 a clear protagonist, and her death stands in for the death of wilderness. When bear 71 is hit by a train, the narrative suggests bear 71's natural instincts are her undoing; in trying to protect her cubs against the oncoming train, she failed to do what does not come naturally – get out of the way. Instead, heeding her instincts, she turned on and roared at the train. Whatever threats a bear and her cubs might face in the 'true' wilderness, they would not be from trains (see Figure 5). The filmmakers' intentions are very explicit: they wanted to make audiences sympathetic to the bear's futile battle against human encroachment. This climax confirms the webdoc's punchline: if the train is the 'leitmotif of modernity', as Leo Marx has argued (1964), then modernity both tempts nature to artifice and sacrifices nature in its forward progress. As Allison said in an interview, 'For me, the train symbolises modernity in a way. [...] There's always this unstoppable force we're having to deal with' (Lederman).

But in my view, the railroad tracks in the webdoc function in more complicated ways. They are not simply a critique of modernity. The fact that the train leaks *grain* is an

Figure 5. Webcam footage of bear 71 and her cubs eating grain on a traintrack.
Source: Bear 71 ©2012 National Film Board of Canada.

important detail, one which is pursued by neither the webdoc nor its reviewers. It is fitting that the scriptwriter of the webdoc's narrative is one of the authors of *100-Mile Diet*, J.B. MacKinnon. MacKinnon and Smith's book is known for arguing that we make our diets more local, so that we reduce our carbon footprint through the transportation of food. His arguments echo through the script in *Bear 71*. Humanity's rapacious appetite for corn grain and corn-fed animals precipitated the mass transportation of corn via railroad, without concern for the material impacts of those transportation networks on the very real places they connect. Bear 71's life and death make real and local what our transportation systems, driven by our appetites, do to nature. These impacts are not calculated in the cost of our industrially produced beef or of our high fructose corn syrup-laden foods. They are externalised as costs elsewhere, most often outside our phenomenological reach; we in the First World often do not perceive them. And the death of bear 71 is a cost, which is all the more invisible to us because we want to see wilderness as 'out there', pristine, untouched by human activity. Those in industrialised capitalist nations like our goods cheap and our nature pure, a contradiction that the webdoc subtly scrutinises. Combined with the knowledge that genetic modification has altered the DNA of corn and monoculture has destroyed ecosystems (background knowledge that is perhaps more imaginable when we remember the scriptwriter's desire for us to reduce our food miles), the notion of a corn-fed grizzly bear population endangered by trains forces us to contend with the naturecultures we have all become.

It also forces us to see that globalisation is not erasing space, as globalisation's critics have argued, but rather that it has very material – and spatial – consequences. This is the essence of the webdoc's potential – to make visual the seemingly placeless – and thereby easy to ignore – costs of human consumption. With closer scrutiny, we may see in the webdoc an argument that increasingly dense transportation and communication networks and technologies only seem to bring places closer together, but that they do not transcend actual places. And we may learn that the binary between wired and wild only serves to remove these impacts from our immediate attention. Globalisation may make it feel like 'there's no there there', but this placeless, messless purity is part of its seductiveness. There *is* a there there, and sometimes there are multiple theres there, as in the kind of networks of places that Doreen Massey (1993) describes when she articulates a 'progressive sense of place' that is simultaneously material and networked, global, and local. *Bear 71* is really not about the bear, then. It's about these circuits of natureculture infoldings, to use Haraway's language (Haraway 2008). It is not enough to simply see the train as an unstoppable, unnatural force against which the bear has no hope, as a literal reading of the webdoc suggests. Rather, a subtler and more sustainable reading of the bear's death on the train tracks, as well as the networked, 'hybrid' spaces of Banff more broadly (as I analyse next), suggests that the bear's death hints at the myriad ways that corn has networked places and blurred lines between nature and artifice, adding grizzly bears to the list of casualties our food miles incur and demanding a better system that internalises the costs of our otherwise out-of-sight, out-of-mind networks of consumption.

Sarah Whatmore's hybrid geography as environmental ethic

Such an insight suggests that the idea of wilderness gets in the way of thinking about these interconnections between the diets of a young urbanite and a grizzly bear, between transportation networks and individual action. Even beloved nature writer and philosopher Gary Snyder (1990) expressed concern about romanticised visions of the wild and advocated for a notion of wildness that should be understood as a quality, not attributed to a particular geography or animal. The environmental historian William Cronon's (1995) essay 'The Trouble with Wilderness' marked a radical shift; he argued that the concept of wilderness poses the greatest challenge to a viable environmentalism in the twenty-first century. Ever since 'Trouble', it is difficult to advocate for that view of wilderness in North American without addressing the problems Cronon outlined: that wilderness is a contemporary idea emerging out of Progressive Era goals of racial and national exclusion; that it involved the forced dispossession of indigenous peoples and the criminalisation of subsistence relationships with the land; and that it is a social construction reflecting more about human politics than ecology.

Many theorists in the wake of this recognition are articulating new forms of environmental ethics that are more inclusive and that use arguments other than pristine 'wildness' to defend nature's value. Some of these alternative definitions are particularly helpful in understanding the implications of *Bear 71*. Donna Haraway defines nature as not a 'physical place to which one can go', but rather as an 'active, changeable presence that is always already in our midst', which 'challenges spatial, as well as social, (dis)positions' (qtd. in Whatmore 2002, 342). Whatmore's notion of a hybrid geography, which draws heavily on Haraway's work, posits that a better environmental ethic would recognise the ways in which the binaries human/wild, nature/culture, animal/human, organic/machine are socially constructed and that recognising the limits imposed by these binaries could help bring forth a new environmental ethic. Citing Cronon, Whatmore argues that 'placing the wild as a pristine exterior harbours within it the very will to power that the gesture would elude' (2007, 12). In place of notions of wilderness that pretend that both the spaces and species of the wild operate outside of human activity, a pretense that occurs at the peril of humans, landscapes, and species, Whatmore suggests a hybrid, relational view 'spun between people and animals, plants and soils, documents and devices in heterogeneous social networks which are performed in and through multiple places and fluid ecologies'. She suggests that this manoeuver occurs by situating ideas about the wild in diverse 'currents and flows' and by 'animating the creatures mobilised in these networks as active subjects in the geographies they help to fashion' (2007, 14). Whatmore's hybrid geography thus conceives 'agency as a relational achievement' involving 'the creative presence of organic beings, technological devices, and discursive codes, as well as people, in the fabrics of everyday living' (2002, 339). Ultimately, Whatmore argues, the 'determination to fix the wild in the geographical and bodily spaces of animals untouched by history is intellectually and practically unsustainable' (2007, 33).

Despite the wilderness tropes in *Bear 71* and Allison's own views of the wild, *Bear 71* articulates the kind of post-wilderness and post-human ethic that Sarah Whatmore and Haraway are espousing. *Bear 71* accomplishes this by situating conventional ideas about the wild within a broader, more dynamic set of relations and by insisting on animal agency. *Bear 71's* emphasis on networks and entanglements may at times be read as a rejection of them, but the form of the webdoc itself ultimately tempers this by registering, at a minimum, ambivalence about the viability of a wilderness 'out there' and showing that we must reckon with some other definition of nature in this globalised, networked world of flows and relations. Even as Allison herself may want to 'fix the wild in the geographical and bodily spaces of animals untouched by history', the webdoc shatters this illusion.

This is revealed in the disconnect between the film's gestures towards the wild, on the one hand, and the ways in which it articulates a hybrid geography, on the other. Allison clearly feels uncomfortable describing webcam depictions of wildlife as disciplinary – that is, as a kind of scientific Big Brother that uses biological information for the purposes of regulatory biopolitics – and claims that she desired to 'get outside' after sifting through hours of footage on her computer. But the technology of webcams, GIS mapping, and interactive multimedia are just as central to the story as bear 71 herself. They create a tension between seeing wildlife as simultaneously only 'data machines' (to use Etienne Benson's term [2010]) and as agents, which challenges the tendency of much wildlife media, such as PBS or BBC nature programming, to fetishise technology as a means of accessing a nature otherwise pure of human imprints. Indeed, the webdoc's own multimedia, blog–videogame–documentary genre-blurring intentions contradict if not undermine its creators' hesitance about whether technology can save nature.

This mismatch between creative intention and the effects of the medium itself make the webdoc endlessly fascinating. The form may have been chosen for its appeal to a target audience, but it (unintentionally) illustrates the conflict that is growing between the assumptions that the mainstream environmental movement holds dear and the reality of the world we live in, where these binaries cannot always hold. By failing to fully engage these debates in the diegesis, the webdoc's creators miss an opportunity to go beyond evoking empathy for the bear or lamenting the loss of the wild. However, the webdoc can be understood as advancing a more nuanced picture of how these hybrid geographies and naturecultures may point to a broader critique of network systems and call for an ethic that accounts for them, rather than engaging in nostalgia for a pre-industrial or even pre-human past.

Centring the user, decentring the wild

One of the important contributions of *Bear 71* is its user-generated platform. As one description of the site notes, 'Participants explore and engage with the world of a female grizzly bear via animal role play, augmented reality, webcams, geolocation tracking, motion sensors, a microsite, social media channels, and a real bear trap in Park City' (Revis 2012). A stalwart environmentalist may find this use of technology *too* user-friendly, *too* catering to the consumer, *too* human-centric. Indeed, because the webdoc at times treats nature as a video game, a virtual world that can be played with, the text is vulnerable to the criticism that it distances humans from nature – the form makes bear 71 a simulacrum of grizzly life, and GIS gridlines a simulacrum of Banff.

But this is the easy critique, and, following Haraway and Whatmore, I propose that the webdoc is making a more interesting intervention. If a problem with other media in similar genres, from nature television to environmental documentaries, is the omniscient perspective of a single narrator, *Bear 71* joins other user-generated new media efforts to put aspects of the extra-diagetical narrative into the user's hands. The effect of this new media genre is that it weakens the overt normative claims of a single narrative by conceding at least some of the narrative control to the user, respecting his or her intention and intelligence. Using only the webcam footage and a minimal story, Allison and the National Film Board of Canada wanted to avoid seeming like a National Geographic feature or a typical nature story, because, as Allison says, 'the material is so unique' (Allison 2012). This text shows how critical *form* is to getting a particular message across. *Bear 71 shows* far more than it *tells*. Allison herself might not agree with what the medium shows, but through the film's use of technologies to create representations of wilderness landscapes and species, we learn that the 'wild' is socially constructed.

As users navigate the grid of Banff, they customise their experiences of the text. The experience feels like a video game in that users are invited into a spatial imaginary of Banff that looks and feels nothing like the so-called real landscape of the place. One website promoting the webdoc emphasises the point that bear 71 moves through an utterly networked landscape: 'the land she travels through has been rendered into a precisely mapped grid, monitored by cameras, sensors, and satellites' (Video, 2012). The gridlines and icons representing train tracks and rub trees are abstract symbols of these landscape features, far removed from a human's lived, embodied experience in that place. This failure to simulate the 'real' thing is intentional, I believe, and has an important effect similar to the narrative's anthropomorphic rendering of bear 71. It reveals the extent to which simulation has always mediated our relationships to nature. Because nature programs, parks, zoos, and even our biology classes teach us about animals and their habitats by simulation, we bring those assumptions and expectations to our experiences of the 'real thing' in, for example, our trips to the wilderness of Banff. Environmental critics have long urged us to recognise the ways in which our views of nature not 'natural', but culturally mediated in all of these ways, from the time we are little children. The lack of verisimilitude between the technologised landscape of *Bear 71* and the landscape of Banff itself makes it impossible for users to transport themselves into the 'real' place in their imaginations. That is, the landscape representation in *Bear 71* is nothing like a virtual reality game like Beyond the Front (2008), a game that invites users into landscape of war-torn Iraq, for which mimesis is absolutely essential for treating post-traumatic stress syndrome. We can worry that the alien topography of *Bear 71* distances users from the landscape, or alternatively we can celebrate the webdoc's suggestion that there is no objective, real, one map of any landscape. The unfamiliarity of the topography reminds users that they cannot imagine how animals perceive landscapes and urges users to denaturalise their accepted aesthetics and perceptions of this landscape.

Second, the unfamiliar topography forces users to challenge their own categories of natural and unnatural, wild and domesticated. Because it does not look like Banff, the gridded landscape of the webdoc makes it hard to delineate between the features that we might want to call natural – like the rub trees the bears and other animals use to communicate to each other – and those we might reject as human-constructed – like train tracks. Symbolically, they all register similarly, blurring the distinction between 'natural' and 'man-made'. Third, the unfamiliar topography forces users to rethink how we understand wilderness spaces. We may want to think of them as pristine and untouched by human activity, when in fact they are profoundly influenced by and materially entwined with human activity – history, politics, laws, and the technologies of communication and transportation. Ignoring these influences cannot help protect nature, and binary approaches often lead to unsustainable views of wildlife and of the intersections of 'human' and 'animal' spaces. *Bear 71's* intentionally unnatural representation of a so-called wilderness landscape reveals to users how unwild even our most precious wild places are. At a minimum, the webdoc shows them as theatres of scientific research, populated with webcams, tagged animals, trails, boundaries, and roads. The 'wilderness' does not exist outside of capitalist apparatuses as material as railroads and as imperceptible as history, laws, and satellite surveillance. That the webdoc is user-guided causes users to see aspects of Banff they may never have imagined, helping to destabilise their perceptions of, and therefore knowledge about, the landscape.

The webdoc's spatial rendering of Banff as a interactive technological grid (rather than a live topo map or navigable photograph, representations with which park visitors and nature lovers may be more familiar) is one of the primary ways that the webdoc contributes to a postmodern and post-human conception of nature, one which provokes us to think about what a hybrid geography of interacting species, spaces, networks, and representations means for us today. It also reminds us how important representations of a space are to telling us how to perceive and value it. We realise the impossibility of ever really understanding how a bear, or any animal for that matter, perceives its landscape, and denaturalises the 'wilderness' as a habitat for wildlife, breaking the association Whatmore described above between wild species and wild landscapes. The animals in *Bear 71* move outside the boundaries of Banff, emphasising the artificiality of any boundary between wilderness and civilisation, showing that relationships between animals and habitats, both within wilderness spaces and outside them, are more complicated than the wilderness myth allows. That is, culture winds through Banff just as the grizzlies wander the alleys of Canmore. Nature is already interrupted and interrupts culture in turn.

The webcam in the wilderness

The role of technology provides a surprising way to arrive at these insights. The technologies featured in this webdoc are webcams, coding that makes the multiplatform text possible, satellites, and users' own access to computers and the Internet. Together, they allow us to view nature in entirely new ways. Technologies that have allowed us to view animals up close and through the life cycle have taught us so much about animal behaviour, and the webcam lets us get us 'closer to' the wildlife we love. Allison talks about 'seeing the landscape where [she] lives in a totally different way' (2012) after looking at the crittercam footage of animals in Banff and around and in the town of Canmore. But she is herself ambivalent about the technology. On the one hand, she says, we are 'losing' 'a kind of subtle knowledge' 'because of technology' (2012), an argument echoed by many environmental writers and thinkers from Edward Abbey and Gary Snyder to Mitchell Thomashow and Robert Michael Pyle. But, Allison also values technology; she thinks that 'these cameras have made us actually value some of these places even more'. After watching 'basically every single day' of bear 71's life, Allison believes she has acquired a familiarity with and appreciation for bear 71 as a creature with its own story, worthy of attention and concern. Because webcams can be triggered remotely, the bears are not scared off by the presence of humans, which Allison thinks allows webcams to 'really capture nature uninterrupted' (2012). Allison reports wanting to get away from her computer so she can go outside and 'really' experience those animals she had been viewing via webcam. But, of course, that is not possible: the whole point is that webcams allow a kind of intimacy that we could not achieve if we were physically present. The irony of 'catching nature uninterrupted' through the highly technologised webcam destabilises the view that humans and human technology are impure while nature is pure. That a camera bothers a bear less than a human would is precisely the point; we may think humans are more 'natural' than cameras, but this is true only if we accept that there is a 'natural' divide between humans and animals, one that is perpetuated by the bear's fight-or-flight instinct, for example.

Further, the webdoc depicts the webcam images of various animals in ways that look like surveillance footage, such as closed-circuit television, or CCTV. Allison reports that Mendes also saw the images this way, even though Allison also reports to have disliked this interpretation. However, at one point in the webdoc, bear 71's narrative tells us that the technology is as much there to protect humans from bears as it is to protect bears from us. We can read this 'surveillance' in *Bear 71* as a critique of the regulatory observation of wildlife. Seeing the footage this way makes Allison empathetically uncomfortable for the same reason that it makes us uncomfortable to not know whether someone is watching us.

As wildlife enters the panopticon, we have to come to terms with the fact that technology is not just producing scientific knowledge. It is becoming a tool of biopolitics. The various dissonances between technology and the wild in the webdoc's content and form, and our attempt to square this with Allison's view, all call for a more complicated sense of technology's ethical potential.

Being Bear 71?

The film tries to overcome these issues by anthropomorphising the bear. As Allison says of watching the footage, 'I started to imagine what bear 71 might be thinking, what might be going through her mind, and so I actually gave her a voice' (Allison). This act of 'giving an animal a voice' is fraught with the ethical and existential concerns of anthropomorphism – projecting human qualities onto animals. No matter how magnanimous the speaker's intentions, speaking for an 'other' is never innocent, and ascribing human intention can never be scientifically sound. The question of how to advocate on the behalf of animals when we cannot presume to know what they think or desire, when even impulses driven by evolutionary biology can be seen as human-created rationales about how nature works, is a moral conundrum that has plagued those who study animals. Anthropomorphising is considered taboo: unscientific, lacking in objectivity, and even potentially destructive to an animal. That Allison would cavalierly approach her material this way is striking. Admittedly, when I first heard Mia Kirschner's sexy voice as bear 71 in the film, I rejected the whole project as crudely manipulative and uncritical.

But it is worth reconsidering the strategy of anthropomorphising in this particular text. As animal theorist, John Berger (2007) argues, there is value in anthropomorphising. Berger argues that until the nineteenth century, 'anthropomorphism was integral to the relation between man [*sic*] and animal and was an expression of their proximity'. 'Today, we live without them', he continues, 'and in this new solitude, anthropomorphism makes us doubly uneasy' (255). Although Berger is more likely thinking about beasts of burden than deer or rabbits, much less any number of communities that continue to live intimately with animals, he is suggesting that arguments against anthropomorphism ignore the ways that it holds the potential to connect humans and animals. Pushing it further, the discomfort with anthropomorphism is not just about scientific objectivity; it is about the pretenses of rationalising nature that underpin the entire project of scientific objectivity in the first place.

But Berger's provocative argument is, in my view, the weakest reason to reconsider anthropomorphising. A better reason to rethink the effects of anthropomorphism in this particular text is related to the text's deconstruction of the binary between human and wild, or, 'bridging' of human and wild. If, as I argue, the film's greatest contribution is its destabilising of the binary between wild and human, then giving the bear a human voice and even a human kind of sentience is perfectly reasonable. Recognising that anthropomorphising is a way to gain knowledge about animals, Frans B. M. de Waall calls for a 'critical' or 'animal-centred anthropomorphising' that abides by the rigors of scientific inquiry.[1] In ethical terms, Joni Adamson (2013) calls for 'disanthropocentric anthropomorphising' along these lines. Stories that involve multiple species across multiple scales and in multiple voices open the possibility for animal agency. Allison's attempt to bring us closer to nature with the bear's story opens up a new way to think about the relationships between humans, technology, and nature vis-à-vis the bear and the need develop an ethic appropriate to that complexity.

Allison, Mendes, and MacKinnon seem to recognise that speaking for the bear in any mimetic way is impossible; the narrative includes information that goes beyond a bear's possible knowledge, both pragmatically and existentially. That is, the storyline, like the spatial representations of the landscape discussed above, does not present itself as 'realistic'. As Allison says,

> we decided to take that leap. [...] [Bear 71] was telling this story of her life as a dead bear looking back. So we decided to put no limits on what she would know. She would be this omniscient bear, she would know what drugs she was shot with when she was collared, she would know about GPS, she would know about everything. But she was also this wild bear, and so she could be this bridge between that wildness and this very short period of time where we've had this incredible explosion of technology. (Allison 2012)

Here, Allison acknowledges that anthropomorphising authorises a radical possibility: that the bear knows about drugs and GPS and can tell her story post mortem. This omniscient position strengthens the film's indirect argument that a grizzly bear – a quintessential emblem of American wilderness – is no less wired than technology is natural. The bear is a bridge.

But all of these 'unbelievable' perspectives allow the creators to invent the bear's view of the loss of the wild in a way that is all the more explicitly defined by humans. The impossibility of knowing a bear's thoughts means that any imagination we have of them is socially constructed. Yet, to acknowledge the social constructedness of a bear's perspective is less anthropocentric than to 'speak for' the bear in a falsely mimetic sense, as some scientific projects like tracking species movements and nature programming do. I see the creators' strategy thus as a form of Adamson's disanthropocentric anthropomorphism. The moral implications of ignoring social construction and presuming to know nature outside of human mediation are perhaps worse than acknowledging that this is a story about humans, not about a bear. This may not have been the intentions of the creators of the webdoc, yet in the end, it is far more ethical in its stance towards the bear than other nature programming, which projects an image of wild that is equally untrue but a lot more seductive, and which hides the role of humans in its construction.

The camera has a witness

The concept of using webcam footage to create a film may make it seem that the bear's story is controlled by the creators, thereby denying the promise of user-generated format of the text. However, I would argue that the bear's voice-over combined with the user-controlled content and the fact that a user's own webcam places the user into the landscape of the game serve a function in that they suggest that 'nature looks back' at us. This is crucial because nature programming, and the way humans have viewed animals in a number of cultural contexts (from entertainment in zoos to edification through scientific observation or as edutainment in nature shows), make animals the object of the human gaze, not subjects able to act in ways that affect us, or subjects that are independent from us and beyond our understanding.

Critical scholarship on the phenomenon of watching animals is illustrative here. Cynthia Chris' book, *Watching Wildlife* (2006), argues that our voyeuristic desire to watch animals in nature programs has been trained over years of colonial and scientific discourse in ways that are now universal through media globalisation. John Berger's famous essay 'Why Look at Animals', from his book turned BBC documentary, 'Ways of

Seeing', argues that looking is always an act of power and that we must interrogate the ways we see for the kinds of oppressions they create or resist. Berger notes, for example, that zoo-goers 'are looking at something that has been rendered absolutely marginal' (260). Through watching animals, animals become 'objects of our ever-extending knowledge. What we know about them is an index of our power, and thus an index of what separates us from them' (257). Randy Malamud rejects zoos as spaces where 'a desire to exert power over' is authorised (2007, 222). Discussing zoos as examples of disciplinary panopticons – the prison structure that Michel Foucault famously argued operates in all kinds of ways in society, not just prisons – Malamud insists that the zoo regulates animals through a one-way surveillance that grants an 'incessant sense of entitlement to absolute spectatorial control' for 'people who relish the power that accompanies total visual access' (223).

Given that these forms of watching animals are forms of power, does this also occur with webcam watching? In *Bear 71*, do we have 'total visual access' to bears, which are merely 'objects of our ever-extending knowledge'? As noted earlier, *Bear 71's* creators are aware of the surveillance aspects of their use of CCTV-style graphics throughout the film. This would seem to make the issue of whether the animals are in the zoo or in the wild almost moot. Indeed, the value of *Bear 71* is to show how this binary between completely caged and completely wild is less rigid than we may have thought.

Donna Haraway's essay 'Crittercam' in *When Species Meet* (2008) explicitly looks at crittercam technologies, arguing that crittercam footage is used to promote 'scientific knowledge', but in fact serves as a far more complex interface between humans and animals. For example, viewers are 'invited to adopt the persona of a videogame player by the semiotic design of the screen'. Further, she contends, 'footage without narration is more like an acid trip than a peephole to reality' (258). But Haraway does not find this to be a misuse of technology or an ethical failure. Rather, she is underlining the point that crittercams are not what they seem – objective, imperceptible, one-way insights into the intimate lives of animals. Like Berger and Malamud, Haraway is concerned with the ability of animals to look back. In these surveillance technologies, what agency do animals have? As Haraway argues, crittercams reveal the impossibility of binaries between nature and culture, human and animal, organic and machine. Crittercams are a 'colonial organism' that illustrate 'the machinic, human, and animal beings whose historically situated infoldings are the flesh of contemporary naturecultures' (261), the 'relentlessly fleshly entanglements of this particular techno-organic world' (262). Haraway sees in these kinds of technologies a recognition in the agency of animals, the ability of animals to look back. The crittercam lets us see how entangled we are with animals, rather than seeing them as 'out there', untouched by culture. These kinds of insights, Haraway concludes, are better than or perhaps more important than the ostensible purposes of public awareness or scientific knowledge that these technologies are deployed to achieve. Rather, recognising the entwinings or infoldings of human and wild, technological and natural, material and discursive or semiotic 'makes us know that situated human beings have epistemological ethical obligations to the animals' in part because 'the animals make demands on the humans and their technologies to precisely the same degree that the humans make demands on the animals' (263).

In *Wired Wilderness*, Etienne Benson (2010) traces the ways that scientists have developed technologies to study animals and historicises technological developments in wildlife biology (from telemetry to radio tagging to motion-sensor cameras) to argue that debates over these technologies reveal much more about how humans contest notions of the wild than they do about the progress of science. Haraway, Whatmore, and Benson are not interested in what we learn from watching animals per se; rather, they take a step back and ask, what does the act of watching animals – whether as scientists, tourists, or media fans – tell us about our ideas of nature? According to Benson, webcams remove the scientist as interlocutor between animal and public, and change people's perception of animal lives in ways that acknowledge, as Berger would put it, 'a proximity' with animals. Removing the 'filter' provided by nature films and their narrators had much more impact on the public than previous forms of observation, such as radio tagging, which was seen, as Benson documents, as a corruption of the wild. In contrast, webcam observations, Benson observes, afford viewers a kind of 'cross-species sympathy' (191). *Wired Wilderness* shows how assumptions about technology and about authentic, 'wild' animal behaviour, assumptions that echo Allison's own ambivalence about webcams, historically have influenced public support for observational technologies.

I cite these scholars' arguments about watching animals because *Bear 71* raises precisely these questions – about the ethics of observing animals and about what the desire to watch animals says about human attitudes about the wild. *Bear 71* achieves what Haraway and Whatmore are arguing for, precisely because it allows for the bear to look and talk back in a variety of ways. A wonderful example in the narration is when bear 71 discusses playing in the river, and irreverently tells the scientists she knows are observing her, and who may be unable to explain her play in biological terms, to 'put *that* in their notebooks'. This comment implicates the biologists behind the webcam, rendering them visible. The narrative hints that the bear may not want to be fully understood by her observers and that she may have motives that humans cannot grasp. The narrative allows for this animal agency without filling in the blanks, again promoting a kind of 'disanthropocentric anthropomorphism'. Another example of this possibility is a section of the story titled 'the camera has a witness', which inverts the subject–object relationship we typically associate with the camera *as* witness. In other words, the technology is being watched by nature. And with our own image embedded within the webdoc's landscape, it is clear that *we* are being watched.

The existence of rub trees throughout the landscape is another form of animal agency; animals are fashioning their environments with their own material-semiotic codes that cannot be 'read' by humans (see Figure 6). Including them in the imagined, virtual landscape of Banff allows the creators to remind us that animals communicate in another world, using their own social media and signs. The prevalence of rub trees in the landscape of *Bear 71* marks a kind of infolding and animal agency. They function as a reminder that humans cannot possibly grasp animal subjectivity, even as we mediate the meaning of animals in culture, and implicate them in our networks and systems. Like volunteer geographic information (VGI), a cutting-edge tool that can challenge dominant representations of spaces such as Google Maps and Google Earth, rub trees challenge human-produced representations of wilderness spaces, what material ecocritics like Jane Bennett (2010) call 'storied matter', which destabilise human authorship. By including rub trees and by not interpreting them, the webdoc further allows animals to create their own landscape outside of human representations and perceptions of it. A biosemiotic reading of the rub trees suggests what Adamson has called 'multispecies ethnographies'; human communication networks are not the only 'voice', so to speak, and so to attempt to 'speak for' bear 71 becomes an important gesture of granting the animal agency. And the form of the text itself – the user-controlled navigation and the bear's narrative, which are both 'unreal' representations of reality – is also part of how this animal agency emerges. That Allison was inspired by Jonathan Harris' (2007) multimedia film *The Whale Hunt*

Figure 6. Rub tree.
Source: Bear 71 ©2012 National Film Board of Canada. All rights reserved.

speaks to the possibility that she wanted *Bear 71* to decentralise the human perspective through new media, just as Harris wanted to 'reach some degree of empathy with the computer' while conveying the story of the centuries-old Inupiat bowhead whale hunt. Similarly, Allison ironically and perhaps sometimes inadvertently uses technology to get humans closer to nature.

Conclusion

In the end, the effect of this multiplatform webdoc is not only to destabilise binaries between wilderness spaces and non-wild spaces, but also between human and animal. The text helps us see nature in ways that account for the multiple levels on which 'nature' is entwined with social activity. Because the webdoc explores how humans interact with animals in an age of networks, satellites, and surveillance, the user comes away from the experience not simply regretting this entanglement as a loss of nature to a globalising and increasingly populated world, but with an acknowledgement that a new ethical orientation is needed precisely because of that entanglement. Traditional notions of the wild – in terms of both *spaces* and *species*, echoing Whatmore – cannot offer a sustainable or viable environmental ethic.

Rather, *Bear 71* redefines how and why we ought to care about nature. The arguments against thinking about wilderness as a social construction and against thinking of animals as impossible to understand are rooted in the fear that if we fail to outline some kind of 'reality' in their wildness, then there will be no ethical grounds on which to argue for their protection. If wildness can no longer be the primary grounds on which to argue for the protection of landscapes and species, what else can we use? I think *Bear 71* suggests an answer, which Allison hints at when she uses the concept of a 'bridge' to describe creating a voice for the bear. That is, to recognise the social constructedness of nature is not necessarily to be anthropocentric, to use animals as mere data machines or sources of voyeuristic pleasure. Rather, to recognise the social constructedness of both wild animals and spaces is to be accountable to them while allowing for their own subjectivity and agency. The webdoc thus constructs a bridge between wild and human spaces and species – a version of Whatmore's 'hybrid geography'. That is, the text asks us to rethink what we think of as wild, both in terms of animals and in terms of landscapes, and, in the way it showcases surveillance and information technologies, posits some kind of entirely new infolding of naturecultures in its place. As Haraway argues about crittercams, *Bear 71* demands a rethinking of our ethical obligation to animals, based on a better sense of the role of technology and media in co-constructing our wild worlds.

Note

1. As described in Wynne (2007).

References

Adamson, J. 2013. "Untitled Roundtable Paper." American Studies Association Conference, Washington, DC, November 23.
Allison, L. 2012. "Necessary Journeys." University of Alaska Southeast. Egan Library, Juneau, Alaska. October 15. One Campus One Book and Evening at Egan Lecture Series Invited Speaker.
Bear 71. 2012. Leanne Allison and Jeremy Mendes. National Film Board of Canada. Interactive documentary. http://bear71.nfb.ca/#/bear71
Being Caribou. 2005. "Leanne Allison." National Film Board of Canada. Documentary http://www.beingcaribou.com/necessaryjourneys/film.html#BC
Bennett, J. 2010. *Vibrant Matter: A Political Ecology of Things*. Durham, NC: Duke University Press.
Benson, E. 2010. *Wired Wilderness: Technologies of Tracking and the Making of Modern Wildlife*. Baltimore, MD: Johns Hopkins University Press.
Berger, J. 2007. "Why Look at Animals?" In *The Animals Reader: The Essential Classic and Contemporary Writings*, edited by L. Kalof, and A. Fitzgerald, Oxford: Berg.
Beyond the Front. 2008. Will Interactive. Video Game.
Blue Planet. 2001. Narrated by David Attenborough and Sigourney Weaver. Discovery Television.
Chris, C. 2006. *Watching Wildlife*. Minneapolis, MN: University of Minnesota Press.
Cronon, W. 1995. "The Trouble with Wilderness." In *Uncommon Ground*, edited by W. Cronon, New York, NY: W.W. Norton.
Haraway, D. 2008. *When Species Meet*. Minneapolis, MN: University of Minnesota Press.
Harris, J. 2007. "Interactive Documentary." *The Whale Hunt*. http://thewhalehunt.org/wh.html
Jacoby, K. 2003. *Crimes against Nature: Squatters Poachers, Thieves, and the Hidden History of American Conservation*. Berkeley: University of California Press.
Lederman, M. 2012. "NFB Bear Doc Highlights Vancouver Festival." *The Globe and Mail*. May 3. http://www.theglobeandmail.com/arts/film/nfb-bear-doc-highlights-vancouver-festival/article4104562/
MacKinnon, J. B. and A. Smith. 2007. *100-Mile Diet: A Year of Local Eating*. Toronto: Random House.

Malamud, R. 2007. "Zoo Spectatorship." In *The Animals Reader: The Essential Classic and Contemporary Writings*, edited by L. Kalof, and A. Fitzgerald. Oxford: Berg.
Marx, L. 1964. *The Machine in the Garden: Technology and the Pastoral Ideal in America*. Oxford: Oxford University Press.
Massey, D. 1993. "Power Geometry and a Progressive Sense of Place." In *Mapping the Futures: Local Cultures, Global Change*, edited by J. Bird, B. Curtis, T. Putnam, and L. Tickner. London: Routledge.
Merchant, C. 2010. *Major Problems in American Environmental History*. Independence, KY: Cengage Learning.
Revis, L. 2012. "4 Inspiring Examples of Digital Storytelling." *Mashable*. January 11. http://mashable.com/2012/01/31/digital-storytelling/
Snyder, G. 1990. *The Practice of the Wild*. New York: North Point Press.
Spence, M. 2000. *Dispossessing the Wilderness: Indian Removal and the Making of the National Parks*. Oxford: Oxford University Press.
"Sundance Institute Announces Artists and Projects for New Frontier at 2012 Sundance Film Festival." 2011. Sundance Institute. October 31. Accessed September 12. http://www.sundance.org/press-center/release/sundance-institute-announces-artists-and-projects-for-new-frontier-at-2012/
Whatmore, S. 2002. *Hybrid Geographies: Nature Cultures Spaces*. London: Sage.
Whatmore, S. 2007. "Hybrid Geographies: Rethinking the 'Human' in Human Geography." In *The Animals Reader: The Essential Classic and Contemporary Writings*, edited by L. Kalof and A. Fitzgerald. Oxford: Berg.
White, R. 1996. *The Organic Machine: The Remaking of the Columbia River*. New York: Hill and Wang.
Wynne, C. 2007. "What are Animals? Why Anthropomorphism is Still not a Scientific Approach to Behavior." *Comparative Cognition and Behavior Reviews* 2: 125–135.
"Video of the Day: Bear 71, An Interactive Documentary from The NFB." 2012. George Stroumboulopoulos Tonight. January 20. http://www.cbc.ca/strombo/arts-and-entertainment/video-of-the-day-bear-71-an-interactive-documentary-from-the-nfb.html

Reading environment(s): digital humanities meets ecocriticism

Stephanie Posthumus and Stéfan Sinclair

Department of Languages, Literatures, Cultures, McGill University, Montréal, Canada

To outline a general framework for a dialogue between the digital humanities and ecocriticism, we begin by exploring the meanings of "reading the environment" and "reading environments" within ecocriticism and the digital humanities, respectively. We then describe a digital ecology where the digital humanities and ecocriticism come together in playful, creative, collaborative interactions. In the second part of the article, we illustrate more specifically the (inter)play of such a digital ecology, by analysing Michel Houellebecq's *The Possibility of an Island*, using *Voyant* (a set of text analysis tools) and *Neatline* (an online suite of tools for plotting time and space in texts). Finally, we explore the broader context of the digital environmental humanities, a new and burgeoning area that still remains largely undefined, but that asserts the importance of the humanities in responding to the ecological crisis while leveraging new tools and technologies.

Introduction: contexts and projects

Framing the relationship between humans and new technologies in terms of whether Google is making us "stupid" (Carr 2008) or "smarter" (Zimmer 2009) makes for good journalism but is not particularly helpful for understanding the complex interactions, identities and systems of the digital age. Binary oppositions tend to polarise the debate and do little to provide historical context and perspective (Wellmon 2012). Rather than reduce technology to the "good," the "bad" or the "ugly," we need to contextualise new technologies in terms of previous shifts in reading and writing, asking how they are different, asking how they define identity and influence autonomy differently and asking how they give rise to different types of interactions between actors and networks. This is the perspective we will adopt in the present article, asking how the digital humanities meet ecocriticism not as an all or nothing engagement, but as a set of give-and-take encounters that open up new forms of inquiry while also remaining faithful to the critical approach of the humanities.

Often wary of digital technologies, ecocriticism has been slow to explore the tools made available by the digital humanities. At the same time, the digital humanities have not yet expressed much interest in questions about nature and environment that are at the heart of ecocriticsm. There are of course a few notable exceptions such as ecocritic Ursula Heise's innovative work with digital humanities specialist Elijah Meeks at the Stanford HUMLab as well as the Rachel Carson Centre's *Environment & Society Portal.*[1]

But encounters between the digital humanities and ecocriticism remain few and far between. As an ecocritic and a digital humanist writing and working collaboratively, we have had several opportunities to try and bring our respective disciplines together (Posthumus and Sinclair 2003, 2011).[2] Yet we have remained keenly aware of the different political perspectives within our fields. Building on our own experiences and an awareness of what are at times irreconcilable differences, we offer the present article as a snapshot of the types of projects that might begin to constitute the digital environmental humanities.

In the first section of the article, we will adopt a more theoretical perspective to discuss reading the environment and reading environments in each of our respective domains, ecocriticism and the digital humanities. Once we have outlined these models and methodologies separately, we will explore the modes of playful, collaborative, creative interaction made possible at their intersection. In the second section of the article, we will focus on one case study in particular, our analysis of French contemporary novelist Michel Houellebecq's *The Possibility of an Island* (2005).[3] First, we will discuss using *Voyant Tools*, a web-based reading and analysis environment that allows the reader/user to play with the literary text, visualise it in different ways and so discover the text as a material entity made up of words, sentences, as well as ideas and themes. Next, we will summarise our exploration of *Neatline*, an online, open-source program that allows the reader/user to create interactive exhibits, visualising time and space in any cultural text, and so explore the environmental imagination understood broadly. It is in these reading environments where interpretation and creation are codependent and where digital tools facilitate new ways of "seeing" the text that researchers and students (re)discover the interactions and processes of digital ecologies.

Ecocriticism: reading (natural) environments

To discuss environments in ecocriticism, an area defined broadly as the study of the relationship between literature and the physical environment (Glotfelty 1996) means opening up a lengthy debate about what constitutes an environment and which environments deserve ecocritical attention. Rather than take a side in this debate, we will examine two images that begin to complicate some of the debate's epistemological and philosophical foundations. The first image is the initial logo of the Association for the Study of Literature and the Environment (ASLE) that shows a human figure sitting under a tree reading a book (see Figure 1).

The idea behind the image is that by going outdoors to read, the literary scholar can begin closing the gap between the written word and the physical world. This has also been the objective behind ecocriticism's early and main focus on non-fiction nature writing that foregrounds encounters with the natural environment. By reading about being in nature or by reading in nature, the literary critic is brought back to the physical world instead of always being led away from it (Buell 1995, 11). This claim has led to much controversy both within and outside of ecocriticism.[4]

We prefer to direct our attention elsewhere to an image that is surprisingly similar to that of the ASLE logo. The Kindle application icon also shows a human figure sitting under a tree but holding something that looks less like a book and more like a tablet.

Reading on a Kindle or any other electronic device outdoors obviously troubles the idea of getting back to nature or even of the "natural" environment. For an ecocritic, the tablet might at first represent the problems of sustainability related to electronic waste, a consumer society driven by new gadgets, the time spent in front of a computer screen

Figure 1. Initial ASLE logo.

rather than hiking or camping outdoors, etc. While these are all valid points that can be made about technology in general, they do not allow us to understand whether the reading environment has actually changed much between the image of the ASLE logo and that of the Kindle application. Do they represent different digital ecologies? If so, how and why are they different? Can we read Thoreau's *Walden* similarly sitting outside with an electronic tablet as we would a printed book? What if we are reading *Walden: A Fluid Text Edition* on the website *Digital Thoreau* that allows users to navigate between all seven manuscripts of the text?

Before responding to these questions, we need to revisit the issue of reading (natural) environments within ecocriticism. For some, the natural environment in the text should be measured against scientific facts known about those ecosystems (Love 2003). When writers get a fact wrong in a non-fiction text, they lead the reader away from a better understanding of the physical world. For others, this reading falls into the trap of the illusion of ecomimesis that glosses over the work of language in creating the world of the text (Morton 2007). And for still others, the scientific facts used to evaluate literary representations of the natural world should also be the object of ecocritical analysis (Heise 1997; Phillips 2003). Whatever the case may be, ecocriticism has worked to create an interdisciplinary approach to literary texts that includes scientific knowledge about nature and the environment. At the same time, it has stuck closely to a model of close reading whether this be in the work of a Romantic deconstructionist scholar like Timothy Morton or a more traditional reader of nature writing like Lawrence Buell. In this sense, ecocriticism models the reading environment illustrated in both the ASLE and Kindle images where a single individual is reading a single book.[5]

This does not change even when the environment being considered in the text is not necessarily the natural environment. Over the years, ecocriticism has branched out "beyond nature writing" (Armbruster and Wallace 2001) to consider many different types of environments – urban, rural, apocalyptic, industrial – in a wide variety of texts – novels, films, video games, plays, to name a few. It has gone from reading the natural environment imagined as a homogenous material whole to reading many different types of environments. Moreover, it has become more critical of the notion of nature from many different theoretical perspectives such as post-colonialism, environmental justice, queer studies and comparative cultural studies to name a few. Yet these studies follow a similar pattern of close reading that involves the usual critical tools of humanistic inquiry even when the sciences, such as evolutionary theory (Carroll 2004), ecological conservation (Elder 1996) and biology (Love 2003), are being called upon to analyse literary representations of (human) nature and the environment.

One final point that needs to be made about reading environments within ecocriticisim is the lack of studies that examine reading as a set of material, cognitive, emotional, physical processes. An early attempt to imagine such processes can be found in William Rueckert's (1978) article, the first in which the term "ecocriticism" appears. For Rueckert, reading poetry can be described in terms of a larger system of energy exchanges where poems act like green plants, revitalising the ecosystem and the reader. Despite its originality, Rueckert's analysis has not been taken up by ecocritics, possibly because it gets caught in the trap of uncritical "metaphorical transfers" from science to literature (Heise 1997). More rigorous scholarship around the idea of ecologies of reading, where "reading" is understood generally as the act of interpretation, can be found in the work of ecomedia scholars.[6] Including the reader or the viewer or the user in the process shifts the focus from the natural to the material in useful ways. Coming back to the ASLE and Kindle images from this perspective, we can see that it is not a question of a more "natural" or a more "artificial" environment; these images represent different ecologies of reading with their own sets of material and physical processes, and possibly cognitive and emotional processes.[7] Such a perspective very much echoes the work being done in ecocomposition on ecologies of writing that considers the writer as part of a system of social processes that constitute an environment (Dobrin 2012). As the current special issue of *Green Letters* attests, these theories of writing can be brought to bear in interesting and productive ways on the question of reading environments, especially when these processes involve new digital technologies.

Going digital: of interfaces and environments

If ecocriticism brings together the individual, the text and the physical environment, with ongoing debate about the informational dynamics between the three parts, reading the environment remains essentially a literal proposition, with a real reader, a stable text and representations of physical environments (even imagined environments retain the literal and dominant sense of environment as the physical surroundings). In contrast, reading environments in the digital context bring to the forefront virtuality and abstraction, with sequences of 0s and 1s resulting in an apparent disappearance of materiality, substance and human intelligibility.[8] There may be a perceived loss; the environment no longer seems to contain the physical text in any stable sense, only ephemeral representations emerge from and disappear back into poorly understand electronic systems. E-readers and other electronic devices are undeniably material, but independently of the more transient digital text.

Still, the virtuality of digital texts does not change how we read literature (individually and sequentially) even if the environments made possible by reading digital texts do change. Reading literary texts sequentially remains an embodied human activity constrained by the mechanics of written language, irrespective of the medium (Hayles 1999).[9] The nature of digital text does, however, have significant implications for literary reading practices. Digital information is composed of discrete units of data, such as a byte that can be used to encode a single letter. Those discrete units can be algorithmically processed and reorganised in infinite ways. It is the digital nature that enables one of the conceptually simplest and most powerful affordances (capabilities) of digital text: search. Depending on the tools available, a reader can search for a single word, a phrase, words in proximity and many other variants. Similarly, digital texts can facilitate what John Unsworth (2000) calls scholarly primitives (basic functions common to scholarly activity) such as discovering, annotating, comparing, referencing, sampling, illustrating and representing. As we reflect on going digital, these activities provoke unprecedented questions about what we can do with texts and what we can do in new digital environments.

Of course, the capabilities of our digital environments are only part of the equation, and we must also consider the capabilities and the inclinations of the human. Not much will be accomplished digitally if a scholar perceives a computer program as antithetical to cherished hermeneutic traditions because the digital text seems unrecognisable, because the user interface seems to force quantitative and positivistic approaches or because the program only seems to produce decontextualised tables of frequency data and undecipherable graphs. These kinds of sceptical perceptions of the computer are widespread in the humanities, in part because few tools are designed with humanities scholarship in mind. Perhaps it does not help that several examples of computer-assisted literary text analysis have sought to use the computer to overcome the soft and fuzzy logic of literary criticism in favour of a more linguistically oriented scientific method. For instance, Susan Hockey (2000), a pioneering digital humanist, argues that computer tools "can provide concrete evidence to support or refute hypotheses or interpretations which have in the past been based on human reading and the somewhat *serendipitous noting of interesting features*" (66, our emphasis). Far from being a mere arbitrator of hypotheses, we argue on the contrary for the rehabilitation of the computer as a powerful engine for producing *serendipitous notings of interesting features*. A digital text can be transformed – even deformed – in ways that produce new representations, new observations and new associations. The construction of arguments based on observed associations is at the very core of the interpretive enterprise, and so digital texts and tools can be seen as powerful allies in the production of new knowledge.[10]

In addition to questions of what can be done with digital texts, there is also a set of questions related to scale: with how many texts can we work? It is revealing that the most common term used for literary scholars working on hundreds, thousands or more texts is "distant reading" (Moretti 1998). The essential argument is that close reading practices are necessarily exclusionary because of the time required for humans to read texts. In contrast, distant reading practices can include an arbitrarily large number of available digital texts so that both the strengths of computer reading (or at least processing) and human reading are leveraged. Prominent examples of this approach (Moretti 2007; Jockers 2013) are most concerned with studying stylistic and topical influence across time, cultures and geography within the context of book history.[11]

But a large corpus is not a requirement for using digital reading environments. Regardless of scale – whether we have a single volume of poetry or tens of thousands of texts from a given century – more can be done with digital texts than with print texts, simply by the virtue of their capacity to be algorithmically processed. The digital is relatively complicated and unstable, and includes a dizzying number of factors such as hardware, software, file formats, character encodings and application interfaces. In contrast, print as a medium is relatively simple and stable – for millennia those have been beneficial traits for access to and preservation of words.

A print book can be considered an interface between the reader and words – it is the point at which components meet and interact (i.e. a human and a text). In a digital context, things are more complex, since the interface might refer to a physical e-reader (similar to a physical book) or to the visual user interface. One can read the same PDF file of a digital text in Adobe Acrobat on Windows, Preview on Mac, Calibre on Linux, GoodReader on an iPad, a Kindle and so on, but in each case, the user interface will determine what can be done with the text. The interface is critical since it is the site of interaction between the reader and the text, from searching to annotating and beyond. For instance, tapping a word on a Kindle to produce a dictionary definition is an example of a user interface that enriches the digital text.[12]

Indeed, it is the inclusion of resources and functionality beyond the text that leads to the need for a term that is more encompassing of a heterogeneous set of interacting components. A reader and an analogue text can be mediated by a print page in a fairly simple system, but the entire infrastructure needed for a reader to access a digital text and any additional functionality is evocative of a larger system that is sometimes referred to as a reading environment. The use of *environment* in a computing context is of course not limited to reading applications but is invoked whenever there is a sufficiently complex system, such as a graphical desktop environment where the operating system might be thought of as the surroundings for any number of desktop applications. Similarly, a reading environment can be the surroundings for a digital text and any number of resources, tools and parameters.

Digital ecologies: learning to play

In "Why Google Isn't Making Us Stupid… or Smart" (2012), Chad Wellmon adopts the expression "digital ecology" as a way of undoing the human vs. technology dichotomy. In his analysis of interpretive practices on the Internet, Wellmon begins by pointing out some of the similarities between footnoting during the Enlightenment and Google's current hypertexual linking processes. Instead of arguing for or against new technologies, Wellmon is asking us to consider new digital ecologies in the light of former ways of reading. Moreover, he remains wary of the dangers of the "algorithmic self" constructed by Google's indexing systems, starting with – though not limited to – the PageRank algorithm that dictates that the more hyperlinks you have referring to your page, the more "popular" you become (and the corollary that if there are no links to your page, you're not indexed, or ranked very low, and therefore for practical intents and purposes your content does not exist). Drawing inspiration from Wellmon's approach, we will develop this notion of digital ecologies in order to explain the ways in which a user engages with digital reading environments.

While Wellmon speaks of "digital ecology or environment," we will use "ecology" as the more encompassing term to include, on the one hand, multiple reading environments, whether on paper or on screen, outside or indoors, and on the other, the multiple interactions made possible by digital analysis tools, whether the reader is scrolling through the text as it might appear in a one-page colour visualisation or clicking on links from one occurrence of a word to another. It is this focus on interactions and processes that will characterise our description of digital ecologies. The computer is not just a scientific tool that can supposedly help prove particular facts about a text. Rather than be preoccupied with the results that can be produced by these tools, we need to focus on the potential richness of the processes that they allow. By thinking more about process than outcome, about multiplying meanings (not data) rather than converging on answers, we can consider how to make these tools an extension of the reading and interpretive practices in which humanists are already engaged.

Experimenting with various ways of producing information leads naturally to a greater emphasis on creativity and even play (while remaining faithful to traditions of interpretation and hermeneutics, rather than, say, distilling data to the most robust set of facts). Stephen Ramsay (2014) refers to this as "screwmeneutics," empowering us to consider our enormous riches of digital texts as a virtual playground where new and compelling knowledge can be produced by trying new things, if we only allow ourselves the freedom to do so. As Malcolm McCullough (1996) states in a manner that is fully compatible with the development of literary interpretation: "play often lacks any immediately obvious aim, other than the pursuit of stimulation, but functions almost instinctively to serve the process of development" (223). This notion of playing with a text nicely complicates the concept of a reading environment where a reader/user engages imaginatively with the text but does not (usually) rearrange the pages or change the format of the book. Making and remaking the text in digital sandboxes, the literary scholar accepts the possibility that she may or may not discover something "new." But the focus shifts to learning about the process of interpretation through the new interactions made possible in digital reading environments.

Bringing the notion of digital play into ecocritical scholarship can lighten some of the gloom and doom scenarios of future life on planet earth. Ecocriticism has largely defined itself as a serious endeavour working arduously alongside environmental activists to reduce human and non-human suffering. Of course, there is a place for laughter and reflection as well as joyous celebration of nature within ecocriticism, but this has not necessarily characterised literary and/or cultural analysis and scholarship.[13] Even early ecocritic Joseph Meeker's (1974) "play ethic" that promotes comedy as an ecological genre, where human behaviour is based on creativity and cooperation rather than on hierarchy and dualism, takes the act of reading the text very seriously. To some extent, digital technologies may be one antidote to reading too seriously, allowing the reader/user to (re)discover not just the joy of reading but also the "pleasure of the text" in ways that Roland Barthes ([1973] 1975) may not have imagined.

Without denying the gravity of the ecological crisis, the reading environments of digital humanities tools invite users to explore other potentialities within the text and within their own imagination. Moreover, the use of these tools often involve the participation of more than one single reader as the literary critic often needs the aid of a digital humanist scholar to navigate the functionalities of the tools. Collaboration requires all of the participants to learn the art of give-and-take so that positions and perspectives are modified through the process. Digital tools do not always "fit" with the texts being analysed and so need to be modified; hypotheses about a text may not lead to any conclusions and so questions need to be rearticulated. As children, we learn to adapt to dynamics within a group and yet as literary scholars we often have trouble with the rigidity of our own ideas. The digital humanities bring to ecocriticism the possibility of new interactions, new collaborations, while ecocriticism brings to digital humanities a different political lens for reading environments.

Putting play into practice: Voyant Tools, Neatline, Michel Houellebecq's *The Possibility of an Island*

If you have read French contemporary novelist Michel Houellebecq's (2005) novel *La Possibilité d'une île* (*The Possibility of an Island*),[14] it may seem like a strange if not mistaken choice for an ecocritical analysis. The novel is extremely critical of environmentalist ethics (the rare times they are mentioned) and it is quite explicit in its portrayal of a post-human world where nature is doing very well. According to the main character, Daniel25, a clone living some 2000 years in the future, the idea of getting back to nature or giving rights to the natural world was extremely short-sighted on the part of the human race. While humans disappeared, unable to adapt to the harsh atmospheric conditions after a nuclear war (this is suggested but never really spelled out in the novel), the living world quickly "reconstituted new equilibriums on top of the ruins of a destroyed world" (Kindle, Loc. 4880 of 5257). While such a prediction of a flourishing natural world might seem less probable in the light of current understandings of possible global temperature increases, it is clear that Houellebecq's novel strongly objects to environmentalist rhetoric about saving nature.

So why begin with this novel if we are interested in forging a path towards the digital environmental humanities? First, the novel has nothing especially environmental about it and yet it poses the question of the effects of very different physical environments. The section of the novel narrated by Daniel1, the human being who lives in a contemporary time period, includes descriptions of urban cities (Madrid, Paris) as well as less-populated areas along the Spanish coastline (Almaria). The section of the novel told by Daniel24 and Daniel25, the 24th and 25th generation of clones of Daniel1, describes first the sterile, isolated environment of the clones (they have no contact with each other except on the Internet) and then the ruins of cities and the regrown forests and lakes of what used to be Europe when Daniel25 leaves his clone residence. Although the clones are no more technologically dependent than we are today, they are limited in their life experiences because of their isolation.

Moreover, the novel also poses the question of reading environments as both Daniel clones are meant to spend their lives reading the "life story" of their human ancestor, Daniel1, and writing commentaries about the meaninglessness of the life of this particular member of the human species, and so reminding themselves of their own superiority as a species that replaced humans. However, in the case of Daniel25, reading his ancestor's life story is what pushes him to abandon his life as a clone commentator and to set out into the outside world. The novel thus poses the question of reading environments in two important ways: (1) by asking how the cognitive and emotional processes of the human and clone narrators differ and (2) by asking how physical landscapes affect the human and clone narrators differently.

Digital text analysis tools: exploring time, thought and emotion

To better understand the cognitive and emotional processes that distinguish Daniel1 from his clone descendants, we began to play with *The Possibility of an Island* in a Web-based text analysis environment called "Voyant Tools" (http://voyant-tools.org). Voyant is deliberately designed to be scalable – it can be used to analyse text from a single paragraph or to analyse text from thousands of documents. The choice of tools and techniques may differ depending on the amount of text, but Voyant is fundamentally designed to facilitate navigation between macro- (distant) and micro (close)-views of a text. The tools augment the reader's analytic capabilities by accomplishing tasks that humans are less proficient at, such as counting, correlating and reconfiguring sequences of words. The main page of

Voyant Tools has minimal content – modelled on a sparse search engine interface – but does specify that it is a "web-based reading and analysis *environment* for digital texts" (voyant-tools.org). The first step in using Voyant is to select an existing corpus (collection of texts) or to create a new corpus by pasting text into the box, or by providing a list of URLs to fetch, or by uploading one or more documents in a variety of possible formats (text, HTML, MS Word, PDF, XML, etc.).

We started with a digital edition of *The Possibility of an Island* that we loaded into Voyant Tools as a single plain text file.[15] This results in a screen with a minimalistic combination of three tools visible (see Figure 2):

(1) **Cirrus** (the word cloud in the upper right-hand corner) is a simple but powerful visualisation that intuitively represents the relative frequency of terms using font size.[16] Note that we have selected to filter out function words that tend not to bear much meaning, such as determiners (the, a, etc.) and prepositions (to, in, etc.)[17]

(2) **Summary** (the bullets in the bottom right-hand corner) provides a very basic overview of the corpus, including the total number of words (118,373) and the number of unique word forms (11,197). The Summary tool is less useful for a corpus with a single document (like this one).

(3) **Corpus Reader** (the panel in the right column) presents a text in a way that it can be read sequentially (text is fetched on-demand as the user scrolls), but also includes enhanced functionality. For instance, hovering over any word produces a display of its frequency (the word "life" appears 239 times) and searching for a word produces the distribution graph along the left side.

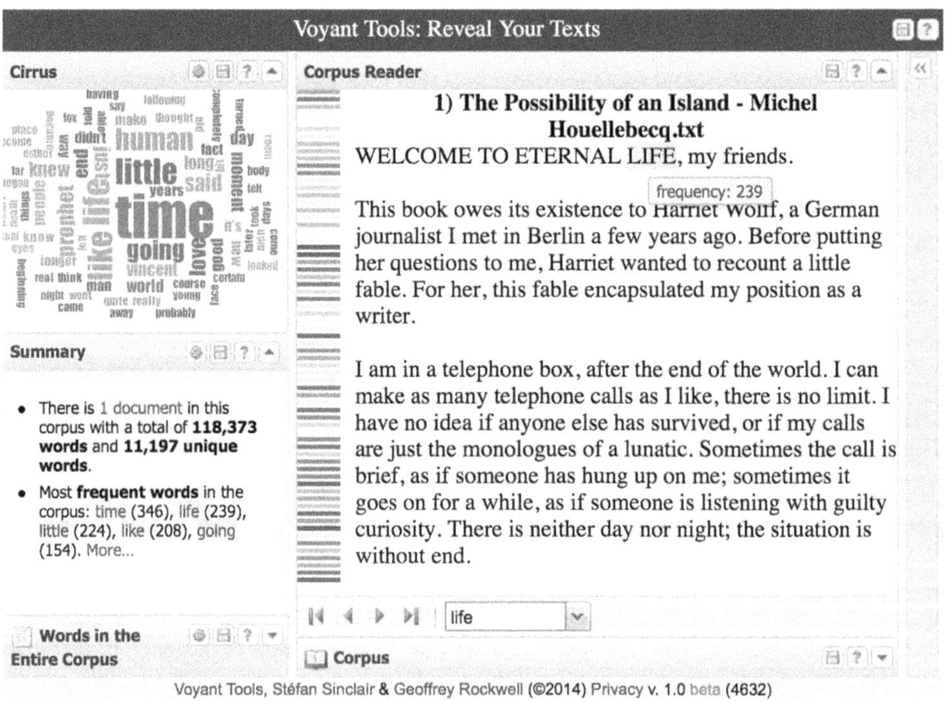

Figure 2. Voyant Tools interface. ©2014, Stéfan Sinclair and Geoffrey Rockwell. Voyant Tools.

It is worth emphasising two very deliberate design characteristics of this initial screen in Voyant Tools. First, the centrality of the sequential text in the Corpus Reader: this is what is most familiar to readers and the interface itself is making an argument about the kind of reading environment that is most useful to literary scholars. Second, the interface is relatively simplified, making visible only three of the eight tools available in the default skin (the default skin is itself a simplified view of the more than 20 interface tools available in Voyant). Additional tools appear through interactions, as when a user clicks on a term in the word cloud. The Voyant Tool developers wished to avoid overwhelming users, especially during their first interactions.

One of the most notable things about the full-text viewed in Voyant is the preponderance of the word "time" in the text. This might be revealing for distant reading (i.e. someone who has *not* read the text closely), but would not be surprising to someone who has read the text with its 2000-year narrative span. What the interface does allow, by virtue of the digital text, is an exploration of the term "time" to better understand how it functions. Clicking on "time" in the word cloud produces a distribution graph of occurrences throughout the text ("Word Trends") that suggests a fairly even spread, with a dip in the middle of the text that might warrant further exploration (see Figure 3). Clicking on any of the bullets in the Word Trends produces a concordance ("Keyword in Context") of the selected term, which allows the reader to appreciate the wide variety of ways in which the word time is used ("the first time," "I had the time," "from time to time," "since the time of," etc.). Thanks to the digital format, the possibilities of exploration are endless.

The Possibility of an Island has essentially three narrators, the human ancestor, Daniel1, and the two Daniel clones, Daniel24 and Daniel25. In our analytic work, we realised that what we really wanted was to have the text from each narrator in a separate document, rather than in the original single document. The digital text is conducive to derivative reorganisations.[18] With three documents instead of just one, the Summary tool becomes much more useful (see Figure 4). For instance, we see the significant discrepancy in text length with a jump of about a factor of about four between Daniel24 (5358 words) and Daniel25 (23,322 words), and then another factor of about four between Daniel25 and Daniel1 (88,925 words). We also observe a shift from Daniel1's interest in people to the later clones' interest in humanity (neo/human/s) and species. Similarly, we see other characters of interest to each clone: the *prophet* (lowercase), *Vincent* and *Esther* for Daniel1, *Marie22* and the *Supreme Sister*[19] for Daniel24, and *Fox* (the cloned dog) and *Marie23* for Daniel25.

Term frequencies in Voyant are indicated for each unique lexical word form such that singular and plural words – and other types of morphological variants – are counted separately, which can be confusing or even misleading. In the example above, combining *human* (57) and *humans* (22) together for Daniel25 produces a value (77) that is greater than the top frequency term: *life* (74). A similar issue arises when we would like to consider all the semantically related terms as a single unit – Voyant does not provide any built-in semantic analysis capabilities. However, it is possible to use external tools (an extended reading environment) such as WordNet to construct sets of related terms that can be combined in Voyant. We did that for the verbs think (http://bit.ly/1eHJ9Xq) and feel (http://bit.ly/1d4FiZH) which led us to hypothesise that Daniel25 is the least thoughtful and Daniel24 the most emotional (see Figure 5).

We have only just scratched the surface of Voyant as a reading environment; this is intended neither as a tutorial of Voyant nor as an in-depth analysis of *The Possibility of an Island*, but rather as a suggestive example of hermeneutic processes enabled by reading digital texts.

Figure 3. Word Trends, Keywords in Context. ©2014, Stéfan Sinclair and Geoffrey Rockwell. Voyant Tools.

Literary cartography: visualising fictional places

To begin analysing the novel's representation of Europe both in the present and 2000 years in the future, we moved from the world of text analysis tools to that of literary cartography. Literary scholars have been interested for a long time in maps that accompany literary texts, especially maps representing imaginary worlds in fantasy novels. As an interpretation of the places described in the story, the map represents another object of study. But literary cartography is not limited to interpreting maps that accompany literary texts; it also extends to the creation of maps based on the geospatial information in or about the novel. Before the emergence of digital humanities tools, such studies were limited to one novel or a small collection of novels.

Franco Moretti's *Atlas of the European Novel* harnesses the power of digital tools to produce a number of different maps, some of which represent places *in* the novels and others, places *of* the novels (their place of publication, translation, etc.). Our study of

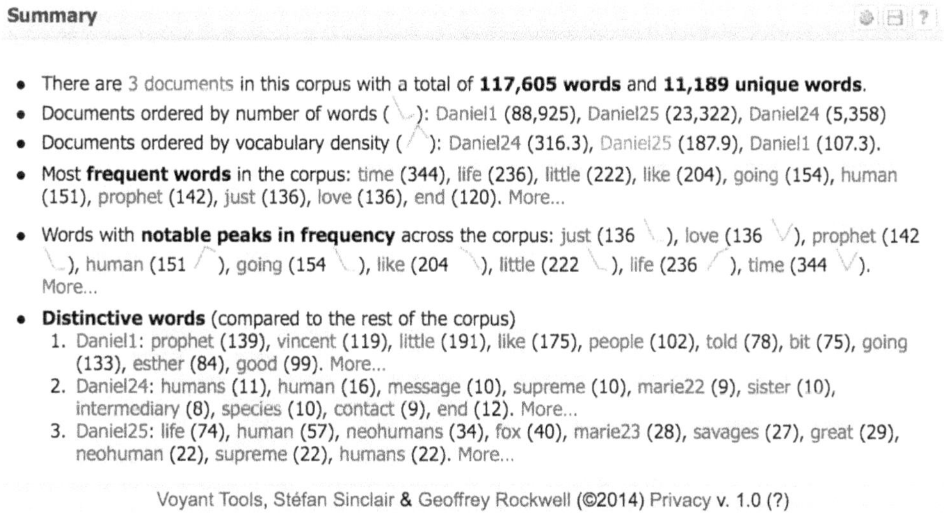

Figure 4. Summary screen. ©2014, Stéfan Sinclair and Geoffrey Rockwell. Voyant Tools.

Figure 5. Cognition and emotion words. ©2014, Stéfan Sinclair and Geoffrey Rockwell. Voyant Tools.

places in one single novel resembles more the first objective of Moretti's work such as his mapping of the towns in Jane Austen's novel. But where Moretti uses rough-grained, less-detailed maps, we were interested in including as much narrative and geographical detail as possible. This is also one of the objectives of Barbara Piatti's large research project "A Literary Atlas of Europe" (www.literaturatlas.eu/en/). Working with a large corpus, this project creates fine-grained maps of places in or across novels. Including more attributes and composed geometries in maps is one way of dealing with the "slipperiness" or uncertainty of cartography that can include fluid geophysical, political, cultural, subjective, imagined and temporal features, among many others (Bushell 2012). Moreover, Piatti's project works on both the close and distant scales so that Berlin is not simply a circle on a map, but a series of differences both in time and in space (for example, Berlin before and after the wall). While such detail obviously closes the distance between the map and the written text, it also requires much more interpretive work on the part of the user and viewer.

In the case of a single novel, creating a fine-grained map that included what was specific to the geospatial information given in the text seemed at first a simple task compared to some of the large data mapping of other digital humanities research projects. Yet we quickly discovered that what was essential to this small project was finding the right visualisation tool. In this case, we chose *Neatline*, an online, open-source digital visualisation tool that allows users to plot time and space in an interactive and easy to use interface. This tool also allows for much creativity, such as uploading hand-drawn, historical or Google maps, and inserting images into the text explanations. We wanted to do more than pinpoint places named in the novel on a set of Google maps; we wanted to show the correspondences between Daniel1's travels and those of his clone Daniel25 far in the future while also illustrating the radical geographical changes.

The final Neatline exhibit is comprised of two maps: one for Daniel1 with small figures and icons indicating places visited on a contemporary Google map of Europe and one for Daniel25, a watercolour map with no geopolitical information but with small figures and images indicating the approximate directions and places of the clone's voyage. The process of mapping the places in the novel led to some interesting discoveries about the passage of time. While the temporal difference between the Daniels' lives may seem striking at first, it is the contrast between the circular time created by Daniel1's visits (he goes back to the same places over and over again) and the more linear time created by Daniel25's adventure that create a space–time connection in the novel. In order to convey this connection, we included lines between the small figures in standing or walking positions that indicate the movement from one geographical place to another in the Daniel25 map (see Figure 6), while the Daniel1 map simply indicates the places visited with still human figures and iconic buildings (see Figure 7).

To add another layer of information to the space–time connection in the novel, we maximised Neatline's other functionalities. The novel describes places down to their actual street names in Daniel1's narrative, so we included an interactive feature for zooming into close-up Google Street views (see Figure 8). For Daniel25's map, we chose images that would give the user/viewer an idea of what such a radically transformed

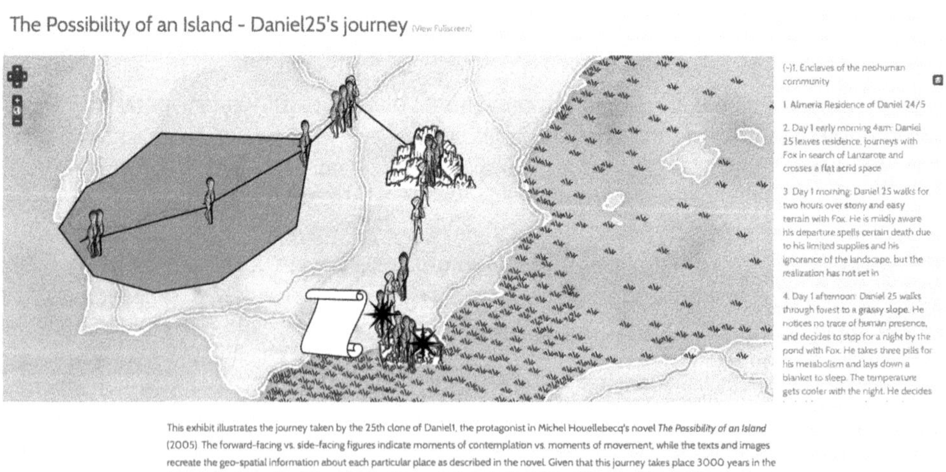

Figure 6. Daniel25 map (open access online exhibit available at http://digihum.mcgill.ca/omeka/neatline/show/the-possibility-of-an-island—daniel25s-journey), © Stamen Design, under a Creative Commons Attribution (CC BY 3.0) license.

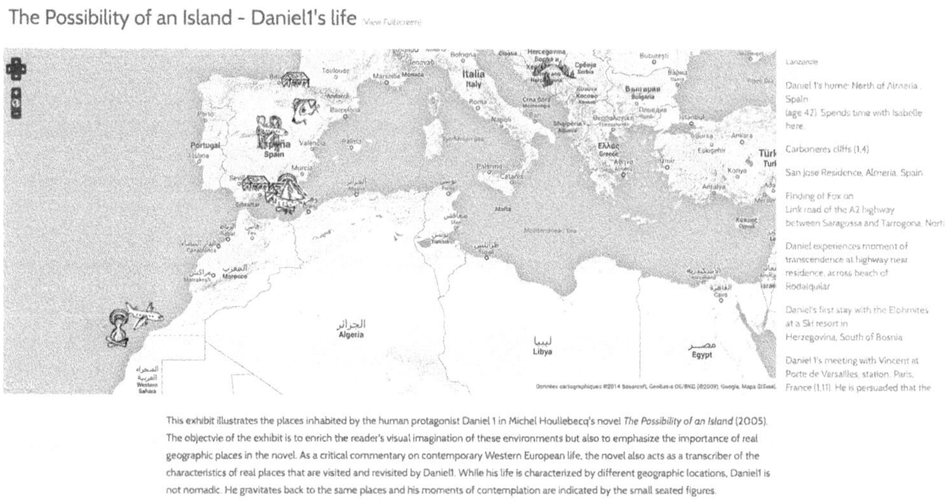

Figure 7. Daniel1 map (open access online exhibit available at http://digihum.mcgill.ca/omeka/neatline/show/the-possibility-of-an-island—daniel1s-life), Imagery ©2014 TerraMetrics, NASA, Map data ©2014 Basarsoft, GeoBasis-DE/BKG (©2009).

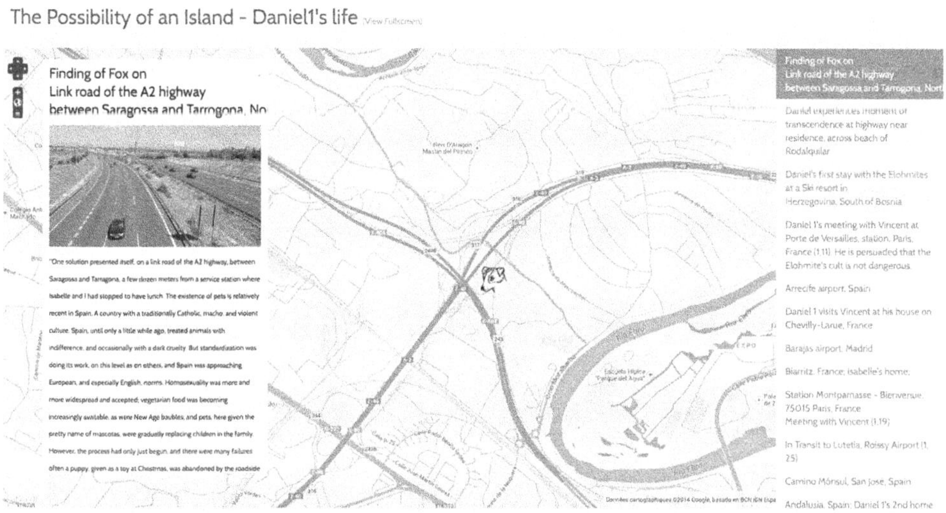

Figure 8. Daniel1 map showing pop up window with Google Street View (open access online exhibit available at http://digihum.mcgill.ca/omeka/neatline/show/the-possibility-of-an-island—daniel1s-life), Imagery ©2014 TerraMetrics, NASA, Map data ©2014 Basarsoft, GeoBasis-DE/BKG (©2009).

landscape might look like (from the apocalyptic, grey dead zone to the lush regrown forests and lakes) (see Figure 9).

Finally, we included short citations from the novel for each of the places pinpointed on the map so that much of the descriptive details were retained (see Figure 10).

In this way, the exhibit represents a hybrid representation of the novel as users can experience the more "distant" view by looking at the map where all places are identified

Figure 9. Daniel25 map showing pop up window with embedded images (open access online exhibit available at http://digihum.mcgill.ca/omeka/neatline/show/the-possibility-of-an-island—daniel25s-journey), © Stamen Design, under a Creative Commons Attribution (CC BY 3.0) license.

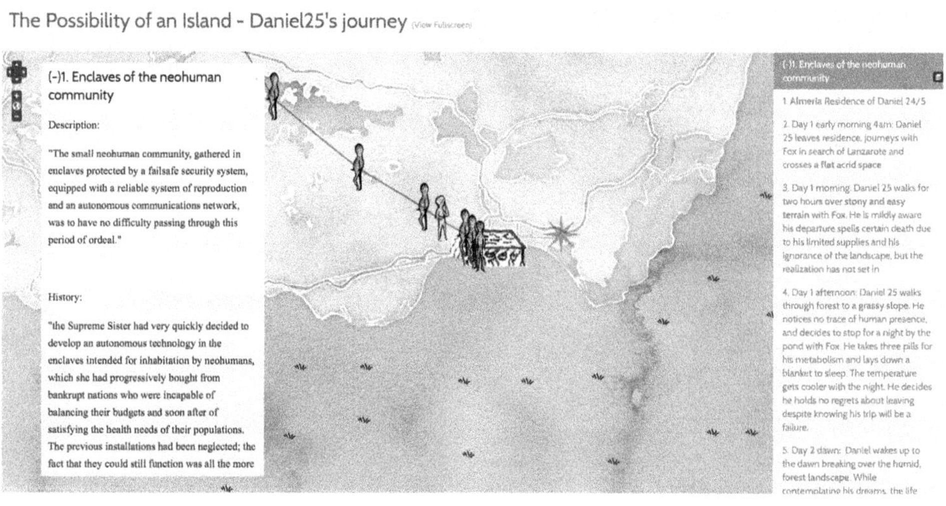

Figure 10. Daniel25 map showing pop up window with citations (open access online exhibit available at http://digihum.mcgill.ca/omeka/neatline/show/the-possibility-of-an-island—daniel25s-journey), © Stamen Design, under a Creative Commons Attribution (CC BY 3.0) license.

simultaneously or they might read through each of the place descriptions individually and experience a "closer" view. Moreover, users can explore the exhibit multiple times and thus experience the text in multiple ways.

While the use of visualisation tools can lead to interesting discoveries about a text, as in this case, the obvious question is: to what extent can this analysis be inscribed into an ecocritical framework? The first response would be that by focussing on the descriptions of place and landscape in the novel, the analysis brings the non-human world into the foreground where it was before in the background. Second, by visualising the spatial information in the text, the analysis draws attention to the processes involved in imagining fictional worlds. Third, by using an online visualisation tool and Google maps, the analysis reveals how our conception of spatiality and textuality is already undergoing important changes.[20] Rather than rejecting such changes, ecocriticism could develop a branch of literary cartography that carefully considers the ways in which maps are used to imagine, discuss and debate environmental issues. This could lead to new interdisciplinary collaboration that shifts ecocriticism away from the natural sciences, like biology and ecology, and more towards the social sciences, like geography and cartography.[21]

Conclusion: digital environmental humanities

Moving from thought/emotion to time/space, from interfaces to environments, from digital humanities to ecocriticism, our snapshot of one single project within the digital environmental humanities may seem less like a composed landscape of carefully placed objects and more a blur of possible colours, features and movements. In this sense, our article captures the heterogeneity that very much characterises the type of work that brings together two very different research areas. But to leave the reader with a broader picture of these questions, we will end by briefly outlining the digital environmental humanities. This is the beginning of what we hope will be a longer and more diverse conversation that includes scholars, teachers, students, journalists and the wider public.[22]

It is important to underline that both the digital humanities and the environmental humanities are working to expand the scope of the humanities as well as assert the value of the humanities in the age of shrinking arts budgets at the university.[23] In the case of the environmental humanities, it is a question of stressing the importance of "a more integrated and conceptually sensitive approach to environmental issues" (Bird Rose et al. 2012), so that the sciences are no longer seen as the sole voice of authority on such questions. To broaden the humanities, work in this area pushes beyond the human, integrating the non-human world in such a way that the boundaries between the human/animal and nature/culture no longer hold. In the case of the digital humanities, it is a question of creating new tools for humanities research that might close the gap between quantitative and qualitative methods. At the same time, this work asserts the importance of a humanistic perspective in making tools and critiquing digital culture. Building on these convergences, the digital environmental humanities will be interdisciplinary in nature and yet rooted in humanistic inquiry; it will bring the tools and methods of the digital humanities to bear on questions being asked by the environmental humanities: how are communities of amateur scientists using the Internet to record and exchange date related to the effects of climate change on their local region? What new possibilities and problems are created when new forms of digital media are used to disseminate research in the environmental humanities? At the same time, it will bring the planetary and political concerns of the environmental humanities to bear on tool making and cultural critique in the digital humanities: how can concepts like the Anthropocene, deep time and speciesism frame the development of new digital reading tools and the use of big data? What does the environmental humanities' emphasis on telling the stories of non-human entities bring to the work of creating digital archives, electronic editions and data visualisations?[24]

It would be naïve to underline the possible convergences between the digital humanities and the environmental humanities without also acknowledging their divergences. One obvious difference is the political imperative within the environmental humanities to establish a more sustainable relationship with the non-human world. The digital humanities may not be particularly interested in critiques of techno-scientific practices that harm the environment. At the same time, the environmental humanities may be sceptical (and rightly so) of methodologies that claim to quantify what are in reality extremely entangled threads of identity politics and power structures. To bring the two together, it may be useful to come back to the idea of digital ecologies. If we adopt Bruno Latour's definition of the digital as materialising and making visible some of the processes of interpreting and reading, we can begin to close what some may see as an unbridgeable gap between the environmental humanities and the digital humanities.[25] Such a way of thinking about digital ecology breaks down the opposition between hardware and software, computer and user, real and virtual, process and result. It asks not how digital reading environments replace reading the environment but how these environments allow for new ways of reading nature and the environment.

Acknowledgements

We would like to thank the anonymous reader for the insightful and extremely useful comments that provided a wonderful springboard for more in-depth reflection and discussion on the possible passages and impasses between the digital humanities and ecocriticism.

Stephanie Posthumus would like to thank her research assistant, Amy Goh, who has worked tirelessly on the Neatline exhibits for over a year and a half, redoing many features and images with each new upgrade of Omeka, the platform on which Neatline runs. Thanks also to Wayne Graham and Bethany Nowviskie both at the University of Virginia for their email responses to multiple queries about Neatline.

Funding

The creation of the Neatline exhibits was supported by a Connections Grant from the Social Sciences and Humanities Research Council of Canada.

Notes

1. Heise and Meek's work using digital analysis tools to mine IUCN Red Lists of Threatened Species has not yet been published, but Heise presented some of the results in her talk "Narrative, Database, Biodiversity Loss" at the Umea University Humlab, 20 September 2011 (http://stream.humlab.umu.se/index.php?streamName=narrativedatabase). The Rachel Carson Centre in Munich, Germany, also has a blog "AntSpiderBee" that posts information about overlapping interests in the digital humanities and the environmental humanities. In the area of environmental history more specifically, the Canadian group NiCHE (Network in Canadian History and the Environment) has led a pioneering effort in the use of digital tools and technologies.
2. We hosted a Digital Environmental Humanities workshop at McGill University in September 2013, which brought together digital humanists and environmental humanities scholars from various disciplines to discuss new projects and collaborations.
3. There are many in-depth analyses of Houellebecq's novel as this author has become increasingly recognised as worthy of scholarly attention after first rising to fame as a best-selling author. It is not our intention here to develop this type of traditional close analysis; instead, we aim to point to some of the possibilities of reading the novel from a different perspective, touching on alternative interpretations and arguing for the importance of these readings alongside of (and not in place of) more traditional close readings.

4. This desire to rejoin the natural world can be traced much further back. For example, romantic poet William Wordsworth invokes readers to get up, leave their books and go out into the "long green fields," to close the "barren leaves" of science and art in order to fully immerge in the "lore which Nature brings" ("The Tables Turned" 1798; http://www.poetryfoundation.org/poem/174826).
5. A quick look at the percentage of single authored articles in ecocriticism's flagship journal, *Interdisciplinary Studies of Literature and the Environment*, reveals that the individual scholar writing about a small corpus remains by far the norm – 87% of scholarly articles from 2013, for example.
6. See for example Adrian Ivakhiv's *Ecologies of the Moving Image. Cinema, Affect, Nature* (2013). Ecocritical film studies seem to have done a better job of focusing on these larger ecologies that include the means of production, the effect on the spectator, etc., than more traditional ecocritical literary studies.
7. Multidisciplinary research on reading in print vs. on screen from the past three decades confirms that the perceived *experience* of reading can vary significantly but that the effective outcomes are relatively similar between the two types of media and depend on a variety of specific contextual factors – see for instance Noyes and Garland (2008).
8. For a sustained though controversial perspective on the perceived losses of print culture and reading practices, see Sven Birkert's *Gutenberg Elegies* (1994).
9. A different argument can be made about reading shorter texts, like articles, and browsing linked and multimedia-rich web pages but we are discussing reading literary novels here. Moreover, we realise that format and medium may have other implications for other aspects of reading culture, such as accessibility and economics.
10. The value of knowledge is a different issue – we make no more claims about the knowledge derived from computational processes than we make about non-computationally derived knowledge. It is useful in this context to recall Umberto Eco's arguments and examples from *Interpretation and Overinterpretation* (1992): humans are association-creating machines. A playful example from the Digital Humanities is Patrick Juola's (2009) Conjecturator which algorithmically produces an infinite number of truth claims about a text, a tiny fraction of which may turn out to be truly interesting.
11. Criticisms of distant reading include the argument that the methodologies represent a disciplinary shift from literary interpretation to quantitative history.
12. Technical journalist Erica Sadun (2014) relates in an article entitled "The Book and I: How the iPad has changed my reading life" how she had become so used to reading on a Kindle that she found herself automatically tapping on a word in a print edition.
13. One notable exception is Michael Branch's creative non-fiction that humorously poses the question of our encounters with nature and culture. In "Are You Serious? A Modest Proposal for Environmental Humor" (2014), Branch argues for the place of the comical and the witty in environmental discourse that often takes itself much too seriously.
14. We are using the Kindle edition of the English translation of Houellebecq's novel (2005).
15. The digital edition cannot be made available for copyright reasons – the legal realities of the digital environment discourage scholarship on contemporary literature.
16. Jonathan Feinberg uses the term "Wordle" for this kind of compact word clouds (see wordle.net).
17. See http://docs.voyant-tools.org/ui/stopwords/ for more information on stopwords in Voyant.
18. Because each section of the text has a title to indicate which Daniel is narrating, it was quite simple to create a script to divide each narrator into separate files.
19. Looking more closely at the occurrences of the term "sister" in context confirms that Daniel1 is referring to sibling relationships, whereas Daniel24 and Daniel25 are referring to the entity that dictates the rules of clone living.
20. As Heise (2008) has astutely noted, Google Earth represents one possible "comprehensive model for considering ecological crisis and environmental as well as cultural connectedness across different spatial scales" (209).
21. French literary critic Bertrand Westphal's *La Géocritique* (2007) is an example of work that draws on geography, cartography and urban studies, to analyse descriptions of European cities in literary texts. Westphal's geocriticism does not, however, use digital technologies to visualise such places nor does it consider the issue of the ecological crisis as fundamentally altering our relationship to space and place.

22. For an idea of the diversity of possible projects in the digital environmental humanities, see the "Projects" page of the website *Digital Environmental Humanities* that is the result of the workshop we organised around this theme in September 2013 at McGill University (http://dig-eh.org/).
23. See the 4humanities website created by a group of scholars in the digital humanities (4humanities.org) and the Environmental Humanities Now website (environmentalhumanitiesnow.org) housed at UCLA.
24. Within the digital humanities, these questions are beginning to emerge (see Bethany Nowviskie's plenary talk at the DH conference 2014; http://nowviskie.org/2014/anthropocene/#more-2445). The environmental humanities have taken so far a largely critical position, examining the problematic use of digital media to represent environmental issues (see the recent issue by Carruth and Marzec [2014]).
25. See Latour's plenary talk at the Digital Humanities conference 2014 in Lausanne, Switzerland (http://dh2014.org/videos/opening-night-bruno-latour/).

References

Armbruster, K., and K. Wallace. 2001. "Introduction: Why Go Beyond Nature Writing, and Where To?" In *Beyond Nature Writing: Expanding the Boundaries of Ecocriticism*, edited by K. Armbruster and K. Wallace, 1–25. Charlottesville: University of Virginia Press.

Barthes, R. [1973] 1975. *The Pleasure of the Text*. Translated by Richard Miller. New York: Hill and Wang.

Bird Rose, D., T. van Dooren, M. Chrulew, S. Cooke, M. Kearnes, and E. O'Gorman. 2012. "Thinking through the Environment, Unsettling the Humanities." *Environmental Humanities* 1 (1): 1–5.

Birkerts, S. 1994. *Gutenberg Elegies: The Fate of Reading in an Electronic Age*. Boston, MA: Faber & Faber.

Branch, M. 2014. "Are You Serious? A Modest Proposal for Environmental Humor." In *Oxford Handbook of Ecocriticism*, edited by G. Garrard, 377–399. London: Oxford University Press.

Buell, L. 1995. *The Environmental Imagination: Thoreau, Nature Writing, and the Formation of American Culture*. Cambridge, MA: Belknap Press of Harvard University Press.

Bushell, S. 2012. "The Slipperiness of Literary Maps: Critical Cartography and Literary Cartography." *Cartographica: The International Journal for Geographic Information and Geovisualization* 47 (3): 149–160. doi:10.3138/carto.47.3.1202.

Carr, N. 2008. "Is Google Making Us Stupid?" *The Atlantic*, July/August. http://www.theatlantic.com/magazine/archive/2008/07/is-google-making-us-stupid/306868/

Carroll, J. 2004. *Literary Darwinism: Evolution, Human Nature, and Literature*. New York: Routledge.

Carruth, A., and Marzec, R., eds. 2014. "Visualizing the Environment." *Public Culture*, February 26.

Dobrin, S. 2012. "Introduction. Ecology and a Future of Writing Studies." In *Ecology, Writing Theory, and New Media: Writing Ecology*, edited by S. Dobrin, 1–23. New York: Routledge.

Eco, U. 1992. *Interpretation and Overinterpretation*. Cambridge: Cambridge University Press.

Elder, J. 1996. *Imagining the Earth: Poetry and the Vision of Nature.* Athens: University of Georgia Press.
Glotfelty, C. 1996. "Introduction." In *The Ecocriticism Reader: Landmarks in Literary Ecology,* edited by C. Glotfelty and H. Fromm, xv–xxxvii. Athens: University of Georgia Press.
Hayles, N. K. 1999. *How We Became: Human Virtual Bodies in Cybernetics, Literature, and Informatics.* Chicago, IL: University of Chicago Press.
Heise, U. 1997. "Science and Ecocriticism." *American Book Review* 18 (5): 4.
Heise, U. 2008. *Sense of Place and Sense of Planet: The Environmental Imagination of the Global.* New York: Oxford University Press.
Hockey, S. 2000. *Electronic Texts in the Humanities.* Oxford: Oxford University Press.
Houellebecq, M. 2005. *The Possibility of an Island.* Translated by Gavin Bowd. New York: Alfred A. Knopf. Kindle Edition.
Ivakhiv, A. 2013. *Ecologies of the Moving Image: Cinema, Affect, Nature.* Waterloo: Wilfrid Laurier University Press.
Jockers, M. 2013. *Macroanalysis: Digital Methods and Literary History.* Chicago: University of Illinois Press.
Juola, P. 2009. "Guessing at the Content of a Million Books." *New Knowledge Environments* 1 (1). http://journals.uvic.ca/index.php/INKE/article/view/154/160
Love, G. 2003. *Practical Ecocriticism: Literature, Biology and the Environment.* Charlottesville: University of Virginia Press.
McCullough, M. 1996. *Abstracting Craft: The Practiced Digital Hand.* Cambridge, MA: MIT Press.
Meeker, J. 1974. *The Comedy of Survival: Literary Ecology and a Play Ethic.* New York: Scribner.
Moretti, F. 1998. *Atlas of the European Novel, 1800–1900.* London: Verso.
Moretti, F. 2007. *Graphs, Maps, Trees: Abstract Models for a Literary History.* London: Verso.
Morton, T. 2007. *Ecology Without Nature: Rethinking Environmental Aesthetics.* Cambridge, MA: Harvard University Press.
Noyes, J. M., and K. J. Garland. 2008. "Computer- vs. Paper-Based Tasks: Are They Equivalent?." *Ergonomics* 51 (9): 1352–1375. doi:10.1080/00140130802170387.
Phillips, D. 2003. *The Truth of Ecology: Nature, Culture, and Literature in America.* Oxford: Oxford University Press.
Posthumus, S., and S. Sinclair. 2003. "Ecology and Technology: 'Dialogue de sourds'?" In *Culture and the State: Landscape and Ecology,* edited by J. Gifford and G. Zezulka-Maillous, 167–177. Edmonton: CRC Studio.
Posthumus, S., and S. Sinclair. 2011. "L'inscription de la nature et de la technologie dans *La possibilité d'une le* de Michel Houellebecq." *Contemporary French & Francophone Studies* 15 (3): 349–356. doi:10.1080/17409292.2011.577620.
Ramsay, S. 2014. "The Hermeneutics of Screwing Around; or What You Do with a Million Books." In *Pastplay: Teaching and Learning History with Technology,* edited by K. Kee, 111–120. Ann Arbor: University of Michigan Press.
Rueckert, W. 1978. "Literature and Ecology: An Experiment in Ecocriticism." *The Iowa Review* 9 (1): 71–86.
Sadun, E. 2014. "The Book and I: How the iPad Has Changed My Reading Life." *TUAW.* http://www.tuaw.com/2014/01/06/the-book-and-i-how-the-ipad-has-changed-my-reading-life/
Sinclair, S., and G. Rockwell. *Voyant Tools.* 15 Oct. 2014 <http://voyant-tools.org/>.
Unsworth, J. 2000. "Scholarly Primitives: What Methods Do Humanities Researchers Have in Common, and How Might Our Tools Reflect This?" *Humanities Computing.* http://people.brandeis.edu/~unsworth/Kings.5-00/primitives.html
Wellmon, C. 2012. "Why Google Isn't Making Us Stupid… or Smart." *ISAC. The Hedgehog Review* 14 (1). http://www.iasc-culture.org/THR/THR_article_2012_Spring_Wellmon.php
Westphal, B. 2007. *La Géocritique, Réel, fiction, espace.* Paris: Minuit.
Zimmer, C. 2009. "How Google Is Making Us Smarter." *Discover.* http://discovermagazine.com/2009/feb/15-how-google-is-making-us-smarter

Signals of nature, prestidigital ecology

Andrew Hageman

Department of English, Luther College, Decorah, USA

This article combines ecocritical readings of three digital texts – the iPad app 'Grass' and Ian Bogost's games *Guru Meditation* and *A Slow Year* – with analytical 'field notes' from teaching these texts in recent EcoMedia seminars. This hybrid approach helps us encounter these digital texts in ways that produce unique insights relevant to ecocritical thought and pedagogy. The article argues that digital texts with overt signals of 'Nature' combine visual perceptions, bodily motions that are simultaneously outside and inside digital spaces, and psychological challenges that leverage memory and anticipation in unique ways that lead us into acts of ecological prestidigitation – sleight of hand. By studying such digital prestidigitation, we can perceive the illusions that have prevented us from thinking and being ecological. Thus liberated, we begin to explore ecology where new ways where algorithms, magic, and imagination meet.

A child said, What is the grass? fetching it to me with full
hands;
How could I answer the child?...I do not know what it
is any more than he.

Walt Whitman (2002)

This article emerges from pedagogical practice and reflection. Over the past three years, first-year students interested in pursing Environmental Studies have taken my four-week January-term EcoMedia seminars. While we have studied a wide range of cinema, television, and musical texts, students responded with noteworthy intensity to the digital applications (apps) and video games that require tactile interaction and deliver haptic encounters. For example, after swiping, rubbing, and tracing their fingers across iPad screens lush with the glowing and gently blowing green grass of the aptly named app 'Grass' (2011), students recorded in blogs and classroom discussions that they experienced vivid childhood memories of outdoor summer play and/or they claimed that they could almost feel and smell the grass. Still other students claimed to be repulsed by what they considered the artificiality of these luminous leaves. Student responses provided two crucial insights. First, digital texts that incorporate tactile interaction, presented in the context of EcoMedia Studies, solicited more textured and multi-sensory reactions than their audio, visual, and audio-visual counterparts. These digital environments fostered student awareness of being in the world rather than sitting back and watching and/or listening to the world. Second, the student responses to 'Grass' as well as other digital texts were often divided into antithetical positions of enchantment and disgust, and yet the grounds upon which they articulated both positions were sometimes the same.

As such, these digital texts helped the class as a whole dig into the ideological roots of our different interpretations of the texts as well as the ideas of Nature and Ecology that we bring to our play.

This article argues for the critical benefits of teaching and analysing human encounters with interactive digital texts, especially those texts with explicit or at least readily visible ecological relevance. Such encounters surpass mere entertainment or escapism, and they do not automatically lead to detachment and alienation from the non-technological world outdoors as Richard Louv claims in *Last Child in the Woods: Saving Our Children from Nature-Deficit Disorder* (2005) as Alenda Chang has argued compellingly (2011). Rather, our active participation in digital constructs that signal Nature help make visible to us the crucial roles that memory, idealisation, and mediation play in shaping both how we conceptualise the world and how we strive to approach the world on its own, non-human terms. Or, put more theoretically, playing with digital apps and video games provides us with unique opportunities for rethinking the paradigms of epistemology and ontology we bring to thinking ecology and to doing ecocriticism. Furthermore, by bringing the digital devices through which we encounter these virtual texts into the classroom, we prompt our students as well as ourselves to analyse the desires for and presumed necessities of having and using such objects, in addition to the texts they deliver. As a result, students become aware of, concentrated upon, and conscientious about our existence as ecotechnical beings – about the choices we make consciously and unconsciously and the ideology that supports these choices.

Regarding methodology and materials, this article combines critique of three primary digital texts with field notes from the EcoMedia classroom. Field notes here include brief descriptions of classroom activities, quoted and paraphrased student response, sorting and categorising of student response, as well as my analyses of student responses. This combination enables us to look at the texts, look at ourselves looking at them, and to draw conclusions that apply to ecocritical speculation and pedagogy. As to materials, in addition to the app 'Grass,' this article explores two video games created by the game designer and professor of media studies and interactive computing, Ian Bogost. The stand-alone game, *Guru Meditation* (2009), features a human avatar meditating while levitating in a blocky 8-bit desert, at least the avatar meditates an levitates so long as the player remains silent, maintains uninterrupted stationary contact with her thumbs on designated areas of the iPad screen, and keeps the device perfectly horizontal. In short, *Guru Meditation* facilitates a haptic experience of oneself physically being with the device while also being inside an estranging, klunkily pixelated environment. The third text is a multimedia piece entitled *A Slow Year* (2010). Utilising the same Atari-based programming parameters that make *Guru Meditation* blocky, Bogost created four games, one for each season. Players must draw upon close observation of each environment, such as trying to catch leaves blown off of a tree in autumn, to score high. Alongside the games, *A Slow Year* includes Bogost's own critical introduction to the game and one thousand season-themed haiku, each 'written' by a computer program that Bogost wrote to 'generate thematically analogous haiku. This generator is able to enforce both the structure of the haiku and the grammatical coherence of the phrases that comprise it' (2010, 21).

All three primary texts synthesise visual perceptions, bodily motions that are simultaneously inside and outside digital spaces, and psychological challenges that leverage memory and anticipation. In other words, these digital texts and their explicit signals of Nature lead us into acts of prestidigitation – sleight of hand. By studying such digital prestidigitation, we perceive the illusions that have prevented us from thinking and being ecological. Yet, we emerge with these insights changed so that we cannot desire an analytical regression to supposedly pristine truths or simple critical decisions between binary terms like virtual and real. So liberated, we can explore ecology as a thing where algorithms, magic, and imagination meet.

What is the Grass?

The iPad app called 'Grass' was produced by CW & T Studio LLC and is free to download via the iTunes store online. When you launch the app, a full-screen patch of glowing green grass appears. Individual leaves constantly yet gently sway as if moved by a light and shifting breeze. Touch the screen and the grass bends in response to your pressure and movements, and if you swipe across the screen in quick repetitive movements, the leaves pass through lighter shades of green to the point of whiteness. Take away your fingers and the original green hue returns as the leaves return to upright postures and resume swaying.

Each time I introduced 'Grass' into the EcoMedia seminar, I checked out from our college library one iPad for each student and brought them to the classroom. Without a lot of introductory set-up regarding the app itself, I joined students in pushing all the desks to the room's perimeter so that we could each find a spot on the floor. After making sure every student had found and launched 'Grass,' I led them through a series of encounter sessions with brief writing sessions between each encounter. Initially, students play with 'Grass' without any additional input for about seven minutes – a span of time that can tellingly feel practically infinite to college-age students playing an app on an iPad that doesn't really do much from a certain point of view. In fact, one of the most consistent points of student response, regardless of how they respond to the more ecological elements of these digital environments, is their surprise at how frustrated they become, at least at first, when they have to do something slowly with this techno-artefact that they almost always use fast and in multiple simultaneous ways. After approximately seven minutes to write about this first play, students return for more 'Grass' sessions accompanied once by Beethoven's *Symphony #6 in F Major*, op. 68 (*Pastoral*) and once by the audio portion of the locally based live-streaming webcam called *Decorah Eagles* (Raptor Resource Project, 2011, http://www.raptorresource.org/falcon_cams/decorah_eagle_xcel.html).

Beethoven's Pastoral is, of course, the music that accompanies the scene in *Soylent Green* (1973), a film we screen in this same seminar, when Sol elects state-assisted suicide and passes away while watching hyper-romanticised footage of deer, sunsets, and waterfalls. The same music and Nature footage reappear as the end credit sequence, and I propose to students a counter-reading of these elements, not as the beauty we need to treasure and preserve, but as part and parcel of the fantasies that enable us to destroy the Earth and eat each other. The *Decorah Eagles* webcam is a chunk of technology that resides in a bald eagle nest approximately four miles from our campus. In its semi-wooded location, this media device channels ambient sound comprised of various birds and other wild animals in the vicinity, the occasional distant bark of a farm dog, and punctuations about every seven minutes of advertisements. In the January 2013 seminar, the ad promoted Buffalo Wild Wings – a bird-enjoyment irony that students immediately noticed.

In this section of the article, I want to focus on three specific student responses to 'Grass' to demonstrate the efficacy of this digital app encounter for revealing fundamental aspects of ecological ideology that typically remain concealed and/or mystified. The first quote presents a thoughtful disappointment with the encounter:

> It's fun to play with for the first few moments, and the gentle waving of the virtual blades of grass inspire a feeling of calm, but then it just starts to feel pointless and frustrating – it won't hold any shape you try to draw. It tries to create the illusion of naturalness, but there's still too much structure. The randomized pattern is still a pattern. The game makes us feel like we can move the grass however we want, and it tries to make us feel like we are really surrounded by grass, that it's under our fingers and we're in control of it, but in the end none of it is real. I'd prefer to go outside and sit on real grass, even if I can't control it the way I can in the app. (Agather, Student Response Writing, 2013)

The overall framework of genuine interest in and openness to this encounter but an overall sense of distaste for the grass of 'Grass' expressed in this response excerpt is not necessarily a surprising result. In fact, this response is representative of more than half of the students who've played with and written about this app in the seminars. What has caught my attention is the intriguing complexity within this largely representative response. This student's disappointment does not arise from a concise binarisation of a real thing and a mediated imitation or shadow of that real thing. The grass of 'Grass' is not filmed footage of living grass. Rather, the digital grass in question exudes both a glow and a pixelated realisation that clearly do not aim at mimesis, or what Timothy Morton, in *Ecology without Nature* (2007), has called ecomimesis: 'Strong ecomimesis purports to evoke the here and now of writing … The reader glimpses the environment rather than the person' (32–33). Put another way, 'Grass' doesn't bring living grass to us so much as it presents self-consciously digital signals of living grass such that a human player encounters specifically digital aesthetic signals of grassiness.

Instead of a concise binarisation, what her response identifies as undermining the encounter is the presence of programmed pattern and structure that fall short of the randomness of an experience with live grass outdoors. This claim raises the question, though, of whether the student actually noticed repetition in the pattern of movements visible on the screen or believed that she recognised repetition by knowing that it must be there somewhere behind the screen. As Kevin Kelly has discussed in *Out of Control* (1994), we are increasingly capable of writing computer programs that convincingly replicate, in animation, the randomness of a colony of bats or a murder of crows flying as a group (10–11). Plus, interactive manipulation sessions of less than 10 minutes each provide a very short time in which to see the limits of this app's randomness. Even if the student was able to catch glimpses of repetition, however, the point here is that the final disappointment with and rejection of 'Grass' is grounded in presuppositions of the limits of technology when juxtaposed with living, organic Nature. The inertia of an eco-ideological notion that only live grass outdoors is real and thereby superior rolled the student through an effective disavowal of the complexities her response registered in order to conclude in a predictable and representative position.

To unpack my assertion, recall the student's two invocations of control. At first, she finds 'Grass' frustrating because the grass won't maintain any images she imprints upon it with her hands. But when she asserts her ultimate disappointment that 'none of it is real,' she underscores her point by saying that she prefers grass outside, 'even if I can't control it the way I can in the app.' The logic slips into contradictions, ever a sign that one is touching upon an ideological core. In this case, the student rejects the digital grass because it does not allow human players to act as Andy Goldsworthy upon it – a desire that already resides in a mixed and sticky territory as it puts together delight in control of the non-human world with acknowledgement of our being in the world as transient and coexistent with others. Then, however, the student claims that the lack of control over living grass outside is a sacrifice contrasted with the digital grass but a sacrifice she welcomes. The contradictions appear in the fact that the digital grass was initially described as frustratingly uncontrollable to the player (despite its being absolutely controlled by the programmer the student knew was present even if undetectable), but in order to recontain the complexities of her response in the comfortable organic-superior-to-technological register, she suddenly posited control in the digital encounter where it wasn't just sentences before.

All of this analysis aims to reopen and prolong the openness of questions this and other students have raised, consciously and unconsciously, in their reactions to 'Grass.' Lest any reader suspect this article is disparaging the students cited, let me be clear that all student responses included here were selected because they reflect rigorous, thoughtful, concerned engagement with ecocritical scholarship and thought. To draw this initial response analysis to a close, consider how effectively a rereading and/or discussion of student responses can open up topics such as what it would mean to control, or be in control of, a patch of grass, digital or in the campus-adjacent prairie. Through the tactile encounter with 'Grass,' we get a prompting to experience and theorise the weirdness of grass and grassiness that is related yet alternate to the Walt Whitman poetic clipping that opens this article or William Wordsworth's 'This Lawn, A Carpet All Alive.' To the multi-textured and alien sensations Timothy Morton ascribes to Wordsworth's poem, 'Grass' delivers similarly uncanny ecological signals through your fingertips or other appendages as they contact the glassy screen (Morton 2004).

Now I want to turn to a pair of student response excerpts that together exemplify the inverse side of the same eco-ideology we have observed in the first sample. One student wrote, 'iPad grass is super fun! It's very realistic, and it's ridiculous how fun it is. The sad thing is, that it is way more fun than playing with real grass. It brings up the point of why preserve the environment when we can create something so cool in its place with technology?' (Thompson, Student Response Writing, January, 2012). Yet another student wrote, 'It is really hard to imagine that I could play with grass in the middle of winter in Decorah, what an experience! I really felt like I was lying outside in my yard while watching the grass sway back and forth. I could almost smell it' (Holloway, Student Response Writing, 2012). Like the first student quoted in the preceding discussion, the latter pair of students tangle with comparisons between living grass and the digital grass they've with which they've just played. However, the latter students did not resolve their complex considerations into a straightforward preference for live outdoor grass. Rather, their writing hangs out longer with the perplexing new way of encountering grass. For them, iPad 'Grass' exists on an experiential continuum of grass and grassiness instead of in an antithetical binary of real and virtual or living and digital grass. While both responses describe sensations of referential memory triggered by the digital grass, the intense memory recalls do not fetishise the past and/or notions of immediate encounters that would make their digital encounters seem lacking. That said, both responses contain ideological curiosities that prevent them from reading like simple *technophilia*.

What catches our attention in the one response is the stain upon this student's enjoyment – the stain of interpellation, as if he suddenly remembered that he needed to temper his enthusiasm, or at least perform a tempering of it. But from where does this guilt arise? Who or what comprises this perceived taboo? The student didn't try to answer such questions in the writing, though this is precisely an issue we took up in the subsequent class meeting. If any one of us is to think and/or say something like, 'I know I shouldn't like this digital thing, but I do,' then we need to try to locate that external authority – that ideology – against which we believe our enjoyment to go. Slavoj Žižek has identified something akin to this problem when he has claimed, 'Even ecology functions as ideology the moment it is evoked as a new Limit: it has every chance of developing into the predominant form of ideology of global capitalism, a new opium for the masses replacing the declining religion, taking over the old latter's fundamental function, that of assuming an unquestionable authority which can impose limits' (2012, 979). This is why I interpret the student's response as interpellation: he felt hailed by a mysterious, and mystified, authority that feels to be beyond knowledge and question, yet this is precisely the area of knowledge and questioning that we need to pursue if we hope to change our ecological thinking and being. And thoughtful encounters with digital ecological texts provide one means for recognising the ideological aspects of ecology and thereby to begin exploring rather than passively accepting them. The student's response even gestures at this claim since his enjoyment-guilt dialectic led him to end with a question, and one that he could not yet answer but felt driven to pursue nevertheless.

Finally, the third student reminded me of William Wordsworth's poem 'I Wandered Lonely as a Cloud' that unfolds a richly textured description of wandering outdoors only to disclose at last that this vision is not being experienced in real-time but is something that plays on the lyrical I's 'inward eye' when he sits pensively inside (line 21). For the student, iPad grass not only felt like real grass, but it almost smelled like it – or, rather, the digital signals triggered olfactory memories. Do we not witness here an inverse of the first student quoted earlier? Just as we must question whether that first student indeed spotted the pattern, we must wonder whether this student indeed felt as if feeling and almost smelling grass? Perhaps the most noteworthy result of wondering this is that for this last student the signals of grass heightened her experience of the present and of memories of live grass encounters. In place of positing binaries and lack, the student used 'Grass' to consider how being human today might mean a multimodal form of being in the world. This strikes me as a productive territory for imagination and exploration: a territory of ecotechnical being and prestidigital ecology where students can begin to join the likes of explorers like J. Hillis Miller (2012) and Douglas Kahn (2013).

Om Mani **iPad***me Om*

This section of the article pivots to Ian Bogost's video game *Guru Meditation*, which is available for 99 cents at the iTunes store. To play this game, you place your thumbs onto images of thumbs that appear on the iPad screen's peripheries, hold the device horizontally level, and be silent. Once you and the machine have achieved an initial steadiness and silence, the virtual yogi on the screen levitates and in the background 8-bit clouds crawl left to right across a desert sky. Too much tilt or a sound as small as a pinky-knuckle cracking sends the yogi crashing back to Earth, game over. In the notes Bogost wrote that accompany the game, he says, 'Time passes subtly during the game. Clouds move. The time of day changes, roughly every hour, from day to dusk to night to dawn' (2009). Meditation through focused yet minimalist interface with a virtual avatar in a digital environment is the object of play. As with 'Grass,' students in the EcoMedia seminars would play the game for a set amount of time, write about their play for a set amount of time, and repeat the sequence of play and writing again. The two student responses quoted and analysed here reveal eco-ideological insights related to yet different from those that became visible via 'Grass.'

A large proportion of students remarked at some point in their written and in-class reactions that 'Guru Meditation' was a radical departure from the video games they play and from what they would even have imagined you could call a video game. While I will not pretend that a substantial number of these remarks were not exactly in service to holding 'Guru Meditation' in high esteem, I will emphasise the power of this game's capacity to jolt us into awareness of the unspoken, unconscious expectations we bring to video games and the devices on which we play them. As an example of students contrasting 'Guru Meditation' with their video game expectations, here is one response excerpt:

> Compared to games like *Halo* and *Mass Effect*, where the fast stimulation makes for a very realistic game, this guru game was boring. However, the game's use of the clouds and mountains was interesting, especially knowing that the game is supposed to put me into a meditative state. Overall I found the game a weird idea of the world in the eyes of technology. Is this how machines see the world we live in? (Plass, Student Response Writing, 2012)

To get at analysis of this quote and its import, consider how, with 'Grass,' players interact with only a tiny swatch of a world. The game, like its name, doesn't signal a world so much as an object extracted from a world. 'Guru Meditation' differs significantly by placing a world before the player. Furthermore, by including a human avatar, the game embeds the player in this other world. While playing 'Guru Meditation,' we are neither exclusively inside nor outside the game. Rather, we inhabit both worlds simultaneously and partially. As a result of this dynamic dual-dimensional encounter, this game opened for my seminarians a space in which we tried to map together the contours of ecological ideology. We recognised that we are always already inside this ideology, but texts that create fundamental dissonance and make visible contradictions seemed to invite us to hover at the border of inside and outside, struggling intellectually along the frontier of what we have and can think about ecology.

The student quoted in the preceding discussion struggled with a double alienation: alienation from video games as a general concept and alienation from a conventional sense of subjectivity as defined by being in the world. The video game alienation was common amongst student responses, and in many of their articulations, this experience resembled a classical *anagnorisis* or sudden realisation of something fundamental that has been going on imperceptibly for some time. College students are part of a generation accustomed to, and by now thoroughly bored by, being warned about harmful video game contents such as the sex and violence of *Grand Theft Auto*. But when they played 'Guru Meditation' in the seminar, the encounter was set within an environment that embraced video games as cultural texts worthy of play as well as analysis and interpretation. And because the contents were so unusual to them, they obtained insights into their standard expectations of the form of video games themselves. The *anagnorisis* I witnessed concerned their own unconscious default expectations that video games mean high speed, multiple actions that require intense concentration of a frenetic, multitasking variety. 'Guru Meditation' demanded not less attentiveness but a radically different variety of it. Not only did students realise this, but the encounter also prompted them to ask questions about how their mode of being when gaming shapes and is shaped by their lives when they're not gaming.

The second element of this student's response concerns subjectivity. The question at the end of the response – 'Is this how machines see the world we live in?' – is another way of asking what type of subjectivity or subject position is the human player of this game being asked to inhabit? What is important is that the tone colours this inquiry as one of neutral, objective intellectual curiosity as opposed to a tone of disgust in the face of a mechanised and/or digitised world. We see an open imagination setting off into a strange new territory of alternate subjectivities, going so far as to entertain the prospect of machines having perspectives of the world. Such a response is striking, though it should also be predictable given that it was prompted by a game that Ian Bogost designed. In his own role in the emerging field of object-oriented ontology, Bogost has repeatedly posited the startling claim that machines, amongst all other objects, experience other objects that are in the world with them. In *Alien Phenomenology* (2012), for example, he provokes the reader to imagine that circuits experience electricity and code just as chiles interact with the other ingredients in enchiladas and human stomach juices: '… how do we as humans strive to understand the relationships between *particular* objects in the world, relations that go on without us, even if we may be their cause, subject, or beneficiary? How do we understand the green chile or the integrated circuit both as things left to themselves and as things interacting with others, us among them?' (29, italics in original). The student's question, like Bogost's theory, displays a critical openness that is not enclosed in either anthropomorphism (the student does not presume that the machine's vision is something we can access fully) or in only object-to-human co-relations (the student's question includes wondering if the game resembles how a machine encounters mountains, clouds, and desert in addition to the yogi).

The second and final student response to 'Guru Meditation' that I include here dovetails with the attention to subjectivity in the first response. This student wrote:

> Ironically, the graphics are far from anything realistic and did nothing to relax me, they actually contributed an irritation factor. The mountains and desert are similar to the technology of old video-games. The clouds drifted so slowly, refusing to let me ignore the annoying landscape. Technology forcing me to slow down to appreciate this artificial scenery was confusing. The guru floating in the middle of the screen gave a subject to the game. It provided a soothing bouncing image to zone out on, but it was only after I closed my eyes, ignored the game's visuals, and existed only in the slight pressure of my thumbs that I felt meditative. (Harrison, Student Response Writing, 2013)

While the previous student got to wondering about machine subjectivity, this student explored the strange double aspect of human subjectivity of playing with and through a pixelated person in a pixelated world.

The description starts panoramically and then zooms in towards the yogi. Finding the clouds painfully slow and the 8-bit landscape irritating, she eventually apprehended the levitating yogi as an entry-point that, in her perhaps unconsciously precise words, 'gave a subject to the game.' The yogi was the figural vehicle that enabled her to transport herself into this uncanny game world, and, while the yogi signals human subjectivity, for someone to identify with this klunkily pixelated person requires a leap of imagination concerning human subjectivity. Significantly, the student no sooner entered the game world through the yogi than she somehow let go of the yogi along with the mountains and the clouds and achieved some manner of meditative state. Even if we want to remain dubious about just how meditative this state was, it is worth noting Bogost's position on this as articulated in the game's notes: 'Naturally, there is an added providence in meditating with one's smartphone, since playing the game necessarily takes the player away from the phone calls, emails, text messages, tweets, and other demands the device usually makes' (2009).

As the student registered her human subjectivity as being constituted by her selves inside and outside the game, her meditation became connected to the low-intensity contact she needed to maintain to keep her yogi-self aloft. Even as she enacting a letting-go of world and human subjective self on certain levels both inside and outside the game, what remained was the undeniable material truth of the digital device in contact with her fingers' flesh. And this truth was made still more vital as we had screened the documentary film *Blood in the Mobile* (2010) earlier in the course, such that all of us in the seminar now looked upon our digital devices somewhat akin to a group of Macbeths, wishing the blood spots out. The encounter became in a sense a meditation on human coexistence with digital artefacts comprised of rare metals, glass, silicon, and delicate circuitry through which electricity transmitted the cloud, mountain, and yogi codes. In other words, the student glimpsed a parallax view of the device and the digital texts it delivers that neither condemned nor embraced the machine and the unsettling subjectivity it enables and makes visible. Such play, such a parallax view, does, however, help us embrace our current material ecological conditions. Through 'Guru Meditation,' we notice that we are stuck in a strange and changing way of interfacing with the world, other objects, and ourselves, and, paradoxically, by accepting that we are stuck do we find a footing from which to move forward out of our current un-ecological mire.

Basho's joystick

Near the start of the book part of *A Slow Year* (2010), Ian Bogost writes, 'Good games, like good poems, are *provocation machines*' (5, italics in original). He asserts a strong connection between video game texts and literary texts, more specifically poetry. Just pages later, he expands upon this connection by claiming, 'Abstraction and condensation make poetry much more similar to videogames than narrative forms like novels and films, an observation that helped inspire the name *game poems* as a subtitle for and description of *A Slow Year*' (12). Bogost asserts this parallel with his eye specifically on video games programmed within the now-primitive-seeming 4k memory parameters of the Atari cartridge's hardware. As a result, the connection may not extend to games driven by complex narratives such as *Bioshock*. Nevertheless, the students in my EcoMedia seminars found Bogost's idea stunning at first, but as a particular set of people who have experience with gaming and literature and who are excited about studying the interdisciplinary zone of environmental studies, they were ready to play and read, think, and discuss, to see if the theory worked out in practice. Overall, students concluded that these games, while once again putting deep challenges to their patience and notions of what video games are and do, did provoke them like a haiku by Basho or Robert Frost's sonnet 'Design,' and quite a few of them remarked that the video games provoke not only their sitting-body imagination to unfold the world in the poem but also their necessary action within this condensed and abstract world. And, while I am focusing here on the games more than on the 'machined haiku,' Bogost's computer-generated poetry challenged students to reconsider how and why they might assert boundaries between human beings and machines as opposed to considering how human beings and machines might co-create new aesthetic dimensions.

In addition to linking video games to poetry, Bogost also links *A Slow Year* to *Guru Meditation*: 'As I reflected on the concept of *A Slow Year*, I realized that I wanted to let this slowness become the game. It would be a game about sedate observation, but one that would embrace gameplay more earnestly than did *Guru Meditation*' (11). The four games of *A Slow Year*, one for each season of the year, engaged the students quite differently than *Guru Meditation*. In place of a single, low-intensity tactile mode of play, now the player repeats actions within the game, moving through a process of rapid learning in initial explorations of the digital environment that comprise a high degree of inaccuracy and failure on the way to increased success and higher scores. As Mark J.P. Wolf has noted in *The Medium of the Video Game* (2001), this steep learning curve was a feature of video games from early on, though initially it was tied to profit-motives of coin-operated arcade machines – people would plug in quarter after quarter to become increasingly deeply familiar with the digital environment and how to act effectively within it (43–46). Since Bogost is not collecting coins at each new game, his games create similar effects in players of a desire to observe and know the environment in practical detail, but to different ends.

The most immediately discernible reactions to Bogost's designed demands for sedate observation and gameplay came from the game 'Autumn.' In class discussions, a number of students said that this game was the one that got them to replay it the most, partly out of their desire to master the game environment and its play objectives and partly because it provoked them to compare the game with their own autumnal outdoor experiences. In 'Autumn,' a tree dominates the screen. Leaves drop from the tree and take irregular paths at varying speeds to the ground. The player moves a line segment left and right in the attempt to catch the leaf before it lands on the ground.

This is far more challenging than it sounds, and a player's success hinges more on the patience and repetition that it would take to get to know any environment deeply than on the usual rapid-fire multitask decision-making that rewards players of games with differently dynamic environments. To play 'Autumn' is to wonder what it would be like to re-play the same game outdoors. What would the difficulty level be if you picked a tree, stood by waiting, and then moved left or right to intercept falling leaves that zig and zag, dip and rise, according to shifting winds and their irregular air-resistance shapes? Students asked questions similar to this one, demonstrating that the game was inspiring them to get to know the non-digital environment outdoors. And my point here is that as ecocritical teachers, and/or parents as well, we should look to digital video game texts like *A Slow Year* as material for a pedagogy that accepts our current techno-cultural conditions as the place in and from which we must find ways to provoke ourselves and others to become ecological thinkers, ecological beings.

Ending in caves

In her astute and truly vanguard essay on video games and ecocriticism, Alenda Chang writes about the famous early game, Will Crowther's *Adventure*. Chang points out that 'unlike most enshrined outdoor activities, caving is less about ascent, panoramic views, and wide open spaces than close confines, restricted vision, and plunging deep below ordinary terrestrial life' (2011, 66). Chang's remark reveals the way we misrecognise much of our ecological ideology as matters of natural fact, without history, purpose, or even prospect of change. But I am arguing that precisely these elements of eco-ideological framework must be perceived, recognised, and actively engaged by us.

Inspired by Chang's piece, I included *Adventure* on the EcoMedia syllabus. The game consists of a black screen backdrop upon which verbal text in glowing green appears. The player receives place-based descriptions and must then type in one- to two-word action commands such as 'Walk East.' Students were assigned a minimum of two hours playtime before the class meeting on *Adventure*. I anticipated resistance if not outright hostility to this game, but the only hostility that manifested in the class meeting was blame for giving them a game that absorbed them and their roommates for much more than the required two hours. In fact, the class meeting swerved into an unplanned direction as we decided to project the game onto the massive screen in the A/V classroom where we met. Two students volunteered to interface while the class put together our collective knowledge from our independent playing in order to explore the digital, text-only cave.

This class meeting was not only a radical surprise to me, but it fostered crucial in-class discussions of the historicity of video games and eco-ideology and of the benefits of re-imagining the binarised dynamics of virtual-and-real, inside-and-outside, truth-and-fiction, and Nature-and-Ecology. *Adventure*, when presented and taught with a spirit of rigour and fun, like all of the digital environment texts included in this article, encouraged students to engage with the prospects of thinking and being ecological in multimodal ways that transcend conventional binaries and all-too-naturalised foundational ideas. Exploring the game with students encouraged me, as well, to critique my own theoretical and pedagogical approaches to ecology and digital texts. Or, put another way, this collective adventure into *Adventure* left all of us collectively working out eco-ideological schemas that did not end up resembling the ready division of Plato's allegorical cave. These EcoMedia seminarians finished the class with a nascent and promising capacity to pursue ecological thought and being the way Werner Herzog's film, *The Cave of Forgotten Dreams*, which we also screened as part of the course, approached the ancient subterranean works of art. We stopped seeking codes to facilitate escapist apathy from or complete understanding of how we currently grasp ecology. Instead, we brought purpose and patience, high-spirits and observation, as well as wonder to imagining more ecological futures through combinations of science, technology, literature, anthropology, magic, mystery, and art. I write this article in hopes that its signals may help us share and remake the work of ecocriticism and ecopedagogy.

Acknowledgment

Special thanks to all of the students who explored EcoMedia with each other and me during J-Terms 2012 and 2013.

References

Blood in the Mobile. 2010. *Dir. Franck Piasecki Poulsen*. Denmark: Koncern TV og Film produktion.

Bogost, I. 2009. *Guru Meditation*. https://itunes.apple.com/us/app/guru-meditation/id306610816?mt=8.

Bogost, I. 2010. *A Slow Year*. Highlands Ranch, CO: Open Texture.

Bogost, I. 2012. *Alien Phenomenology, or What It's Like to be a Thing*. Minneapolis: University of Minnesota Press.

The Cave of Forgotten Dreams. 2011. Dir. Werner Herzog. MPI Home Video.

Chang, A. 2011. "Games as Environmental Texts." *Qui Parle: Critical Humanities and Social Sciences* 19 (2): 57–84.

Grass. 2011. CW & T. https://itunes.apple.com/us/app/grass/id417630059?mt=8.

Kahn, D. 2013. *Earth Sound, Earth Signal: Energies and Earth Magnitude in the Arts*. Berkeley: University of California Press.

Kelly, K. 1994. *Out of Control: The New Biology of Machines, Social Systems, and the Economic World*. Reading, MA: Addison-Wesley Publishing Company.
Louv, R. 2005. *Last Child in the Woods: Saving Our Children from Nature-Deficit Disorder*. Chapel Hill, NC: Algonquin Books.
Miller, J. 2012. "Ecotechnics." In *Telemorphosis: Theory in the Era of Climate Change*, edited by Cohen, T., Vol. 1, 65–103. Ann Arbor, MI: Open Humanities Press.
Morton, T. 2004. "Wordsworth Digs the Lawn." *European Romantic Review* 15 (2): 317–327. March. doi:10.1080/1050958042000l680642.
Morton, T. 2007. *Ecology without Nature: Rethinking Environmental Aesthetics*. Cambridge, MA: Harvard University Press.
Soylent Green. 1973. Dir. Richard Fleischer. Beverly Hills, CA: Metro-Goldwyn-Mayer.
Whitman, W. 2002. *Leaves of Grass and Other Writings*. New York: Norton.
Wolf, M., ed. 2001. *The Medium of the Video Game*. Austin: University of Texas Press.
Žižek, S. 2012. *Less Than Nothing: Hegel and the Shadow of Dialectical Materialism*. New York: Verso.

The sustainability of our digital environments: the language of the upgrade path and e-waste

Caroline Stone Short

Department of English, University of Florida, Gainesville, USA

This article looks at the material consequences of our use of digital technologies that are becoming obsolete at an increasingly rapid pace. I argue that how we write about digital technologies has an effect on the technologies themselves and that scholars who study digital forms have a responsibility to take these material consequences into account.

The digital's increasing popularity may provoke us to ask new questions about interactions in 'virtual worlds'. Digital environments – whether they are physical spaces 'augmented' by apps, massively multiplayer game worlds or virtual reality applications used for therapy and entertainment – merit serious scholarly attention. These interfaces are changing health care, pedagogy, play and even warfare. As Gregory Ulmer (2003, 29) argues about the shift from literacy to what he terms 'electracy', new technological developments demand that we invent new methods of scholarship and teaching. Ulmer says, '[t]he goal is not to adapt digital technology to literacy … but to discover and create an institution and its practices capable of supporting the full potential of the new technology' (29).

As we continue to theorise about and invent scholarly methodologies suited to digital environments, we should take care not to limit our focus to the software-level affective dimensions of these technologies. In other words, in our consideration of emergent digital environments, it is important to keep in mind that these spaces are never separate from the other environments that belong to what M. Jimmie Killingsworth (2010, 88) has called our 'lifeworld'. The ecological sustainability of the technologies that make these electronically mediated experiences possible, the material consequences of digital devices, these questions have the potential to bridge what may appear to be a divide between the physical world and digital environments.

The question of speed

On the one hand, the sheer speed of technological change frustrates our attempts to slow down and notice of the material consequences of the digital. In 1965, semiconductor researcher and co-founder of Intel Gordon E. Moore predicted in an article for *Electronics* that the 'future of integrated electronics is the future of electronics itself' (Moore 1965, 115). In the article, Moore anticipated that the processing power of integrated electronics would continue to increase 'at a rate of roughly a factor of two per year'. Although Moore's prediction was for the near future or 'for at least 10 years' (115), this rate has remained roughly the same since his paper was published. Now, some 50 years later, we know that for the most part, Moore's Law has correctly predicted that 'semiconductor power would double every eighteen months to two years' (Grossman 2006, 144).

This doubling of semiconductor power leads to a high rate of turnover of electronic devices, as the faster processors outpace and eventually render obsolete the previous year's machines. Or, as Jennifer Gabrys says, 'While reference is often made to the quickening of information through digital technologies, it is evident that speed has a necessarily material, cultural, and chemical composition' (2011, 28). This digital upgrade path moves so quickly that cell phone carriers Verizon, T-Mobile and AT&T announced in the summer of 2013 that they would be offering financing plans that allow customers to make monthly payments on their existing phones and to upgrade their devices every 6 to 12 months (Gryta 2013).

The devices released today are already on their way towards obsolescence. And, this technological obsolescence is planned far in advance. Killingsworth argues that the most significant consequences of a techno-rhetoric that 'breeds forgetfulness of geographic places and neglects the experience of bodily presence' (2010, 88) are effects this 'tendency to forget' has on the lifeworld. He says:

> The images of landfills teeming with plastics and metals from discarded computers should raise big questions about software giants who render our equipment obsolete with every new version of their products. Should we really need to replace faculty workstations every three years, as the current wisdom at my institutions suggests[?] … What is the environmental cost of such planned obsolescence? (2010, 88)

This 'techno-rhetoric' goes hand in hand with the planned obsolescence of the upgrade path, each fuelling the other, as processor power continues to double.

The material remains of digital environments

When computers cease to function with the rapidity that is required to keep pace today, they frequently end up on shelves, in drawers, in closets. In fact, '[t]here are now so many pieces of abandoned high-tech electronics stashed in private basements and closets that some recycling professionals refer to them as "closet fill"' (Grossman 2006, 140). There is no recent comprehensive data on the amount closet fill stockpiled in US homes and offices (Saphores et al. 2009, 3324), but a 2009 study conducted by researchers at the University of California, Irvine, estimated that each US home has roughly 4 small (less than 10 pounds) and 2.4 large e-waste items stockpiled (Saphores et al. 2009, 3323). The researchers estimated that there are more than 700 million e-waste items stored in homes, a figure much higher than the Environmental Protection Agency's estimate, which does not account for larger electronic items such as refrigerators or ovens (2009, 3330).

When consumers do choose to recycle their electronics, most cities do not accept e-waste through kurbside collection, and these items must be taken to local collection centres or retailers that offer electronics recycling bins. In 2012, the Institute of Scrap Recycling Industries estimated that between three and four million tons of electronics were processed in the US in 2011 (2012, 40). After these items have been de-manufactured at local facilities, different components such as cathode ray tubes often must be shipped to other states that have facilities certified to dismantle hazardous components. Furthermore, '[a] great deal of electronics are dismantled for materials recovery by US federal, state, and county prison industries' (Grossman 2006, 249). And the US is the only industrialised country in which inmates are forced to process e-waste (Grossman 2006, 254). One of the partnerships to receive the most attention is Dell's use of UNICOR (Federal Prison Industries), the federal prison system's recycling business. Inmates at the penitentiary in Atwater, California, recycle Dell's computers. A report by the Silicon Valley Toxics Coalition found that 'manually smashing leaded glass in cathode ray tubes unnecessarily exposed [these] workers to risk of toxic contamination and cuts' (Davis and Smith 2003, 4). In 2002, Occupational Safety and Health Administration (OSHA) found that lead and cadmium levels in the air at the Atwater facility far exceeded the permissible exposure levels (Grossman 2006, 251). Although all of UNICOR's seven facilities obtained responsible recycling certification in 2011 and are now OSHA certified, this only came after a 2010 Office of the Inspector General investigation found that UNICOR failed to protect workers from hazardous substances at its recycling centres (US Department of Justice Office of the Inspector General 2010). As Elisabeth Grossman argues in *High Tech Trash: Digital Devices, Hidden Toxins and Human Health*, the use of prisoners who are paid less than one dollar a day to recycle electronics 'poses serious questions about labor practices and business competition' (2006, 249).

When electronics are recycled there is no federal legislation in the United States against the exportation of e-waste to countries that illegally import this waste in order to mine the discarded items for materials that are of value (Ogunseitan et al. 2009, 670). The Basel Action Network and Greenpeace have worked to draw attention to how this problem is affecting the global third world where workers often disassemble electronics by hand, exposing themselves to dangerous chemicals. In Guiyu, China, which CNN has called 'the electronic wastebasket of the world' (Watson 2013), electronic waste is processed at an alarming rate, often through 'informal' e-waste collection channels. According to a 2012 United Nations University report on e-waste in China, 'informal recycling focuses on extracting re-use and scrap values from e-waste without environmental protection measures, emissions controls or measures to protect the health and safety of workers' (Wang et al. 2013, 4). Although there is 'data scarcity' for figures on informal e-waste processing in China, the report concludes that these 'activities have caused serious damage to local environments and the health of workers in such locations as Guiyu in Guangdong Province and Taizhou in Zhejiang Province' (Wang et al. 2013, 4). And the groundwater, since the mid-1990s, has been poisoned by the run-off from electronics recycling and must now have water trucked in (Grossman 2006, 185). Grossman (2006) describes the Guiyu electronics workers' conditions of labour:

> In small outdoor work areas, workers sit in front of small braziers where circuit boards containing plastics and sometimes fiberglass, impregnated with chemicals and metals, are melted down. When the boards are soft and liquid, the solder and chips are plucked from the toxic soup. The lead-based solder is collected for metals dealers, and the microchips are transferred to another area of Guiyu for further processing. There, beside a river, chips are bathed in open buckets and barrels of acid to extract the gold in an ages-old process called 'aqua-regia' ... The process uses a mixture of pure nitric and hydrochloric acids to precipitate the gold out of the plastic and other metals. Puffs of caustic steam – gas likely to contain chlorine and sulfur dioxide – rise from the liquid's surface. After the gold has been reclaimed, the waste liquids – pure acids and metals – are routinely dumped in the river. (187)

And on the production side, the fabrication of microchips and the fabrication of the printed circuit boards on which they will eventually reside both have significant environmental impacts. According to Eric D. Williams, an environmental researcher who has written extensively on the impact of microchips, '[t]he semiconductor industry uses hundreds, even thousands, of chemicals, many of them toxic' (2004, 2). Furthermore, the production of microchips requires substantial amounts of energy and natural resources including water and silicon (2004, 2). As Grossman (2006) notes

> one square centimeter of finished silicon wafer weighs about 0.16 grams – or about half the weight of a hummingbird egg, which is about the size of a small jelly bean (but considerably lighter). … making this small and incredibly delicate item requires about twenty liters of water. (42)

To take just one plant as an example, the world's largest semiconductor manufacturing plant, Taiwan Semiconductor Manufacturing Company (TSMC), uses millions of gallons of ultra-pure water a day to produce its chips. And the plant can only expect business to increase. Since 2011, TSMC has partnered with Apple to produce chips for the next generation of iPads and iPhones. Although it took several years for the company to produce chips that met Apple's demands; in June 2013 Apple announced that it would be purchasing chips from the plant beginning in 2014 (Lessin, Luk, and Osawa 2013). And while TSMC has offered classes in water recycling to other businesses in the area and the plant recycles some of its own wastewater, the sustainability of this plant's consumption is questionable, especially in a region that is experiencing ongoing drought conditions (Williams 2004, 4). Similarly, the fossil fuels used to fabricate a chip, 'total 600 times the mass of the final product' (Williams 2004, 5). The emissions from the plants that produce microchips in developing nations are often not monitored (Williams 2004, 3), but there have been recorded cases of water contamination from these plants. And the workers who produce the chips are at a greater risk for health complications. The chemicals workers are exposed to in semiconductor wafer fabrication might be 'explosive, toxic, or corrosive' (Koh, Chan, and Yap 2004, 180). Multiple studies have shown that these workers' reproductive health has suffered, and that the women in the plants have a higher rate of miscarriage than the general population. And former workers at IBM and National Semiconducter have filed lawsuits alleging that they experienced higher rates of birth defects and cancers (Williams 2004, 4). While goods powered by microchips are lauded for their abilities to improve the quality of our health – we now have e-doctors, heart and blood sugar monitoring iPhone apps, and ways to 'gamify' recovery from illness and trauma (McGonigal 2012) – this focus on only the promises of digital devices overlooks those workers whose health has suffered from the manufacturing of these goods.

A problem of language

On the one hand, it is possible to place the blame for the vast material consequences of digital environments on the speed of the technological upgrade path and planned obsolescence. But this perspective, I would argue, limits scholars' agency to work towards change. Instead, I would suggest that certain language used to describe digital environments has also worked towards writing over these consequences of electronic interfaces. The language we often use to describe these technologies – specifically, here, the phrases 'the virtual' and 'new media' – furthers a tendency to forget their effects. Although I would argue that this terminological problem extends far beyond these two phrases, their prevalence in scholarship about digital environments and the theme of this issue of *Green Letters* on 'virtual worlds' both make these phrases fitting places to begin.

With respect to electronic technologies, 'the virtual' is now used as a broad categorical term that perhaps conveniently, but inaccurately, is used to speak about a wide range of devices and practices. From writing about massively multiplayer digital environments such as *Second Life* and *World of Warcraft* to the US's ongoing programs of drone warfare in Afghanistan, Pakistan and Yemen, authors invoke 'the virtual' to describe image-rich electronically mediated interfaces. By labelling these technologies as virtual, a gesture is made to a difference between reality, materiality and actuality. A virtual thing, in this context, is 'almost, but not quite'. Discussions of simulation and illusion are closely related to these invocations of the virtual. My primary concern about the pervasiveness of 'the virtual' in writings about digital media stems from my belief that this language calls attention away from the material consequences of the technologies that are labelled as 'virtual'.

Even when scholars are attentive to the material consequences of electronic devices, this reliance on the language of the virtual has a tendency to persist. In these cases, the language works counter to the authors' stated goals. For example, in *Climate Change and the Media*, Tammy Boyce and Justin Lewis rightly argue that we should become more aware of the environmental effects of the telecommunications industry. Yet, for all their emphasis on the materiality of forms of media, in their description of the industry's content, Boyce and Lewis say, 'The virtual nature of much of the industry's content tends to obscure their responsibility for a vast proliferation of hardware, all with high levels of built-in obsolescence and decreasing levels of energy efficiency' (2009, 5). I believe, as Boyce and Lewis argue, that it is important to hold the telecommunications industry responsible for the environmental impacts of digital media, but, for me, central to holding these companies responsible is not separating the hardware from the software. As Friedrich Kittler argues in *Literature, Media, Information Systems*, there 'is no software at all' (1997, 150) because at some level the software is entirely dependent upon and bound up with the hardware of a machine. Or, as Gabrys argues in *Digital Rubbish*, reworking Kittler's formulation, '[I]n programming matter, software becomes tied to matter; it constitutes a distinct articulation of material processes. … There is no software because there is nothing soft – or absent – about it' (2011, 62). In other words, my argument is based on the 'all the way down' materiality of electronic devices. From the content to the marketing campaigns to the shipping materials, a thorough media ecological study takes into account the material effects of the many aspects of these technologies, including the content. We cannot blame the 'virtual nature' of content for obscuring this responsibility. Rather, the responsibility is ours as well, and one of the ways that we must become accountable is to recognise this linguistic contradiction. The content has never been virtual in the sense that it has been less real; it has only been perceived, described and accepted as simulated and dematerialised. Although computer interfaces are often referred to as virtual, their material consequences remain.

Similarly, the phrase 'new media' directs our attention away from the influence of obsolete forms of technology. And while the use of 'new media' persists since its rise to popularity in the mid-1990s, increasingly there are signs that scholars recognise that this phrase does not fit with technologies that rapidly become obsolete, that are not perpetually new. For instance, in their essays in Charles Acland's (2007) collection *Residual Media*, both Jonathan Sterne (2007) and Lisa Parks (2007) suggest that this continued use of 'new media' is problematic. Sterne (2007) says, '[O]n the scale of media history, computers have been around for quite some time, yet we persist in calling them "new media." The question is why' (19). In part, then my attempt to trace the proliferation and popularization of this phrase in academic writing about digital forms of media is my initial attempt at answering Sterne's question. Or, as Gabrys says, 'There is certain impossibility to writing the "now" of these technologies, a fact that is only made more evident by the strange persistence of the self-defeating term *new media*' (2011, 114). From Gabrys' perspective, 'new media' is self-defeating because as soon as new digital technologies emerge, they are already on their way to quickly becoming obsolete. In this sense, then, all forms of new media might be thought of as old new media.

Self-defeating as it may be, scholars are in the habit of using the phrase and are sometimes even willing to perform verbal gymnastics to continue saying 'new media'. Paul Levinson addresses this terminological problem by referring to old new media as 'classic new media' (2009, 1) and then goes on to define the forms that have most recently emerged as 'new new media' (2009, 1). He says:

> New new media are so new that few of them had a major place in our world five years ago. Several did not exist four years ago. … But few of these new new media are discussed in classrooms at any length or detail in textbooks nor in many other kinds of books, either. (Levinson 2009, 1)

Although Levinson rightly calls attention to the lack of scholarship that surrounds some of these newer forms such as 'Twitter, Second Life and podcasts' (2009, 1), his work may also suggest the urgency for questioning what happens to 'classic new media' or old new media. What will happen in another 5 years – or sooner considering, again, the question of speed – to 'new new media'? Will these forms be referred to as 'classic' or will we call the forms that have emerged most recently '*new* new new media'? Levinson's book *New New Media* serves as one example of how commonplace the use of 'new media' has become, but the adherence to the phrase 'new media' is widespread. Used as a general category to describe computerised technologies, it is impossible for this phrase to be accurate. The 'new' cannot remain perpetually precise as a label. But, the problem with this phrase goes beyond its inaccuracy. This problem with this persistence of a rhetoric of the 'new' is that it directs our attention.

In other words, the phrase 'printing press' might bring to mind hands, ink, maybe damp newsprint. Perhaps, the 'printing press' invokes Benjamin Franklin's wooden press, or perhaps it calls to mind a more recent iteration of the printing press – a three-story machine of metal, towering rolls of paper and the whirring of gears as it rapidly prints thousands of pages in an hour's time. But what about 'new media'? Although many have attempted to define what falls into the category of 'new media', from the outset this phrase precludes the old. So, it is not enough to define 'new media' as electronic forms of media because that would include the dusty, outdated machines in a university's storage closets and surplus warehouse, waiting to be auctioned or scrapped. 'New media' implies that digital technologies are always the newest of their kind, the most recent upgrades.

The wall that now exists between 'old' and 'new' media is an example of how divisions create unintended remainders. The phrase 'new media' creates a zone of ambiguity, a categorical void into which the forms previously referred to as new media fall when they have been rendered obsolete by the upgrade path. These old forms of new media can no longer be considered as properly new. They are no longer the most recent versions of electronic devices; yet, because this division between 'old media' and 'new media' attaches a specifically mechanical and analogue meaning to 'old media', these outdated digital forms are not considered as properly old – this is the liminal status of old new media.

Tracing discursive consequences

If we perpetuate these divides between 'virtual' worlds and the physical environment and forms of old and new media, we are complicit in directing attention away from the material consequences of technologies. How we write about these technologies is one of the factors that affects the lives and deaths of the devices. 'New media' obscures digital forms of media, with increasingly shorter and shorter shelf lives. This phrase is the linguistic equivalent of packing them away in storage closets, out of sight, forgotten. Not only does the phrase 'new media' elide older electronic technologies, but it is also ideologically aligned with the upgrade path. As Sterne says in 'Out with the Trash'

> In a weird, recursive way, new media are 'new' primarily with reference to themselves. Computer culture has reached a truly bizarre equilibrium. Today, computers and other digital hardware displace their own counterparts more than anything else. 'Newness' in computers is defined with primarily reference to old computers. Along with cell phones, they are designed to become obsolete after a short period of use. They are designed to be trash, to make room for future profits, additional hardware sales, and performance upgrades. Certainly, computers have become a vehicle through which users can encounter 'the new' in media technology. (2007, 19)

Sterne (2007) believes that 'we need digital hardware that is more democratic, slower to change, easier to use, and less damaging to the environment' (29). Although I do not disagree with the Sterne here, I would argue that what we also need is language and methods that do not cover over the residual influence of discarded electronic objects. Although 'the virtual' and 'new media' work to 'write over' the histories of outdated digital technologies, the effects of these devices remain substantial.

Connected through digital environments

Digital environments may seem separate from 'real' places, but the creation of 'virtual worlds' has consequences that are far from simulated. Being mindful of these consequences can begin with changing how we write about the digital. When we think about the digital not as separate but a powerful part of our environment, as the term 'digital environments' encourages us to do, we work against the prevailing tendency to separate electronic technologies from their ecological consequences. These questions of separation and connection are not uncommon to conversations about digital devices; however, I would suggest that material consequences have the potential to reframe some current conversations about our interactions and interfaces.

While much of early writing about 'new media' lauded the powers of computer-based communication to network us, make us interact, bring us closer together, there is an increasing sense, today, that we have gotten so close to our tiny screens, so immersed within interfaces, that now we are each alone with our devices. This is Sherry Turkle's argument about digital technology. But she is not alone in feeling alone. Novelist Jonathan Safran Foer's (2013) editorial for *The New York Times* 'How Not to be Alone' echoes Turkle's sentiments in her 2012 TED talk and book *Alone Together* (2012). Foer's piece, which was adapted from a commencement address he delivered at Middlebury College, describes his ambivalence at seeing a young girl on a bench in New York crying:

> Technology celebrates connectedness, but encourages retreat. The phone didn't make me avoid the human connection, but it did make ignoring her easier in that moment, and more likely, by comfortably encouraging me to forget my choice to do so. My daily use of technological communication has been shaping me into someone more likely to forget others.

The 'others' here that Turkle and Foer worry that we are forgetting are the others around us who are also using hand-held digital devices. Foer and Turkle both seem to be saying that it is possible to be disconnected from these others, and if we do, we are alone.

But there is another sense in which we have never been 'alone', and our digital devices only further this connection we have to others. This is not a connection newly achieved through digital environments; it is a connection through global resources and the labourers who work for international corporations. This difference between being separated from or connected to others depends on perspective, on the angle of the lens we choose to employ. It depends on whether or not certain ways of being connected matter to us. In a 2013 performance, comedian Louis C.K. characterises this oscillation we experience between selfish, solipsistic thinking and our willingness to consider the bigger picture as the 'of course, but maybe' moment (2013).

> Everybody has a competition in their brain of good thoughts and bad thoughts, hopefully the good thoughts win. For me, I always have both. I have the thing I believe, the good thing, that's the thing I believe, and then there's *this* this, and I don't believe it, but it is there. It's always this thing, and then *this* thing. It's become a category in my brain that I call, 'of course, but maybe'.

C.K. walks his audience through how the 'of course, but maybe' principle works, moving from the somewhat commonplace and mundane to increasingly serious examples. 'Of course', he says, children with peanut allergies should be protected, and 'of course' soldiers are heroes. 'Of course' slavery is bad, 'but maybe … there's no end to what you can do when [you don't care] about particular people. You can do anything. That's where human greatness comes from, we're shitty people'. At this point, C.K. takes his iPhone out of his pocket.

> Even today, how do we have this amazing microtechnology? Because the factory where they're making these, they jump off the … roof because it's a nightmare in there. You really have a choice. You could have candles and horses and be a little kinder to each other or let someone suffer immeasurably far away just so you can leave a mean comment on Youtube.

The 'but maybe' that we so infrequently consider is the answer, I believe, to Turkle's and Foer's sense that, sadly, we are alone with our devices. Or, from a different angle, Turkle and Foer are correct. We have forgotten others while we have satisfied this urge to get closer and closer to our screens. But maybe the others we have forgotten are the nonhumans and the humans that are influenced, moved, changed and put into service by, but certainly not left alone by our reliance on digital technology.

Moving forward and glancing backward

Scholars in the humanities are beginning to turn their attention to electronic waste. In *Technological Ecologies and Sustainability*, editors Danielle Nicole DeVoss, Heidi McKee and Richard Selfe devote the final section of their collection to a consideration of these technologies' material consequences 'beyond our borders' (13). The editors describe e-waste as an issue that 'has been too long neglected in our field' (13). In Devoss's introduction to this section of the book, she argues that our discussions of sustainability should never be limited to ways to sustain scholarship alone. Rather, we have to consider 'the ways in which technologies leave trails, and where tools go to die, or, ideally, to be reborn' (DeVoss 2009, 1). As Apostel and Apostel say in their chapter of this collection, 'Old World Successes and New World Challenges: Reducing the Computer Waste Stream in the United States', in these 'various developing countries … [the items] are picked apart by hand, exposing impoverished workers to the hazardous components inside' (2009, 3). They describe the situation as 'a computer waste stream crisis' (Apostel and Apostel 2009, 2). What is remarkable about this chapter goes beyond its content, which is similar in many ways to Grossman's and Gabrys' work. That this chapter appears in a recent collection with other well-known scholars in rhetoric and composition suggests a growing recognition within the discipline that e-waste is 'one of the most critical sustainability-related issues we face' (Apostel and Apostel 2009, 2).

But raising these issues within the humanities is only half of the battle. Assuming e-waste and the sustainability of digital technologies are subjects that will continue to gain traction within the liberal arts, the task becomes how to create more sustainable practices of scholarship. But simply calling for newer methods risks falling into the neophilic trap of believing that all we need to fix the current problem is newer and newer objects. Rather, I would argue that we already have critical approaches that can be applied to electronic sustainability. Just as, after Latour (2004) and others, we have begun to approach science as constructed, we need to approach the digital upgrade path as created, in part, through our own language. Although some have argued that objects have been overlooked for too long at the expense of studying 'language, discourse, culture, and values' (Coole and Frost 2010, 3), approaching media studies through rhetoric does not mean that we must abandon the things themselves as powerful and 'vibrant' (Bennett 2010, 241). Objects are influential actants that do not exist solely as linguistic constructions. On the other hand, how we write objects has afforded them part of their influence. The task, then, is not just to understand how 'our writing … our research designs, our consumption practices, our sympathies' shift when 'this "call" from things is taken seriously' (Bennett 2010, 240) but also to understand that our scholarship has changed matter itself, in part through directing attention to what should matter.

To take one example of how studying the language of the upgrade path changes these possibilities for action, we can look at how language changes something as seemingly concrete as Moore's Law. In a 2012 *Wall Street Journal* article about microchip manufacturing, writer Shara Tibken begins by describing the challenges Moore's Law faces. She cedes agency to Moore's Law, describing it as a precarious thing that might not be able to keep pace – 'Moore's Law … faces a major hurdle' (2012). What is especially interesting is how Tibken's description of Moore's Law varies from how writers in scientific discourse communities describe Moore's Law. In these accounts, too, Moore's Law is given agency, but it is the agency ceded to other well-accepted scientific principles – Moore's Law determines, the law predicts, the law holds that, and so on. What is even more curious is how Tibken describes Moore's Law at the end of the article. This time she places the semiconductor researchers in control – '[s]till, the investments to date show the industry is placing most of its bets on EUV as a contender to keep Moore's Law going' (2012). And here, writing to an audience whose primary lens is about financial markets, companies' quarterly profits, the 'how' of the upgrade path becomes clearer. Moore's Law, of course, is not as fixed as the law of gravity. Rather, our acceptance and belief that this rate of technological turnover would continue is what has made it so. The language, here, matters. Either we are helpless to push back against the rapid obsolescence of electronic devices, or what has caused this rapid obsolescence of electronic devices has been, in part, how we have written and therefore conceived of and bought into technological change. This is why it matters whether we use 'new media' and 'virtual' to describe digital environments. Alongside studies that take as their subjects the cast-aside devices of the upgrade path, we need materialist accounts of the field of media studies itself. Situating this backward-looking analysis of media studies methods and language alongside how the fields' objects of study have circulated within networks of influence figures media scholars as active participants in the shaping of objects' travels and interactions. And, when what we find is that media objects leave behind them trails of toxic waste and sickened workers, an archaeological account of digital media studies positions us as scholars to become accountable for some of the ways we have furthered these consequences.

References

Acland, Charles R. ed. 2007. *Residual Media*. Minneapolis: University of Minnesota Press.

Apostel, Shawn, and Kristi Apostel. 2009. "Old World Successes and New World Challenges: Reducing the Computer Waste Stream in America." In *Technological Ecologies and Sustainability*, edited by Daniel le Nicole DeVoss, Heidi A. McKee, and Richard Selfe. Logan, UT: Computers and Compositions Digital Press.

Bennett, Jane. 2010. *Vibrant Matter: A Political Ecology of Things*. Durham, NC: Duke University Press.

Boyce, Tammy, and Justin Lewis. 2009. *Climate Change and the Media*. New York: Peter Lang.

CK Louis. 2013. *Oh My God*. YouTube.

Coole, Diana, and Samantha Frost, eds. 2010. *New Materialisms: Ontology, Agency, and Politics*. Durham, NC: Duke University Press.

Davis, Sheila, and Ted Smith. 2003. *Corporate Strategies for Electronics Recycling: A Tale of Two Systems*. San Francisco, CA: The Silicon Valley Toxics Coalition.

DeVoss, Danielle. 2009. "Introduction to Section IV." *In Technological Ecologies and Sustainability*. Logan, UT: Computers and Compositions Digital Press.

Foer, Jonathan Safran. 2013. "How Not to Be Alone." *The New York Times*, June 8.

Gabrys, Jennifer. 2011. *Digital Rubbish: A Natural History of Electronics*. Ann Arbor: University of Michigan Press.

Grossman, Elisabeth. 2006. *High Tech Trash: Digital Devices, Hidden Toxins and Human Health*. Chicago, IL: Island Press.

Gryta, Thomas. 2013. "Verizon to Offer More Frequent Phone Upgrades if Subscribers Forgo Subsidies." *The Wall Street Journal*, July 19.

Institute of Scrap Recycling Industries. 2012. *ISRI Scrap Yearbook 2012*. Washington, DC: Institute of Scrap Recycling Industries.

Killingsworth, M. Jimmie. 2010. "The Body in Eco-Rhetoric and Techno-Rhetoric." In *Rhetorics and Technologies. New Directions in Writing and Communication*, edited by Stuart Selber, 77–93. Columbia: University of South Carolina.

Kittler, Friedrich A. 1997. *Literature, Media, Information Systems*. New York: Routledge.

Koh, D., G. Chan, and E. Yap. 2004. "World at Work: The Electronics Industry." *Occupational and Environmental Medicine* 61 (2): 180–183. doi:10.1136/oem.2002.001073.

Latour, Bruno. 2004. "Matters of Fact/Matters of Concern." In *Things*, edited by Bill Brown. Chicago, IL: University of Chicago Press.

Lessin, Jessica, Lorraine Luk, and Juro Osawa. 2013. "Apple Finds it Difficult to Divorce from Samsung." *The Wall Street Journal*, July 1.

Levinson, Paul. 2009. *New New Media*. London: Pearson.
McGonigal, Jane. 2012. "The Game That Can Give You 10 Extra Years of Life." TED.com. https://www.ted.com/talks/jane_mcgonigal_the_game_that_can_give_ you_10_extra_years_of_life
Moore, Gordon. 1965. "Cramming More Components Onto Integrated Circuits." *Electronics* 38 (8): 114–117.
Ogunseitan, Oladele, Julie Schoenung, Jean-Daniel Saphores, and Andrew Shapiro. 2009. "The Electronics Revolution: From E-Wonderland to E-Wasteland." *Science*, October 30.
Parks, Lisa. 2007. "Falling Apart: Electronics Salvaging and the Global Economy." In *Residual Media*, edited by Charles R. Acland, 32–47. Minneapolis: University of Minnesota Press.
Saphores, Jean-Daniel M., Hilary Nixon, O. A. Ogunseitan, and A. A. Shapiro. 2009. "How Much E-Waste is there in U.S. Basements and Attics? Results from a National Survey." *Journal of Environmental Management*. 90: 3322–3331. doi:10.1016/j.jenvman.2009.05.008.
Sterne, Jonathan. 2007. "Out with the Trash: On the Future of New Media." In *Residual Media*, edited by Charles R. Acland, 16–31. Minneapolis: University of Minnesota Press.
Tibken, Shara. 2012. "Smartphones Challenge Chip Limits." *The Wall Street Journal*, August 27.
Turkle, Sherry. 2012. "The Flight from Conversation." *The New York Times*, April 23.
Ulmer, Gregory L. 2003. *Internet Invention: From Literacy to Electracy*. New York: Longman.
US Department of Justice Office of the Inspector General. 2010. *A Review of Federal Prison Industries' Electronic-Waste Recycling Program*. U.S. Department of Justice Office of the Inspector General. http://www.justice.gov/oig/reports/BOP/o1010.pdf.
Wang, Feng, Ruediger Kuehr, Daniel Alquist, and Jinhui Li. 2013. *E-Waste in China: A Country Report*. United Nations University. http://isp.unu.edu/publications/scycle/files/ewaste-in-china.pdf.
Watson, Ivan. 2013. "China: The Electronic Wastebasket of the World." *CNN*, May 30.
Williams, E. D. 2004. "Environmental Impacts of Microchip Manufacture." *Thin Solid Films* 461: 2–6. doi:10.1016/j.tsf.2004.02.049.

Index

Note: **Boldface** page numbers refer to figures and tables. Page numbers followed with "n" refer to footnotes.

For Product Safety Concerns and Information please contact our EU
representative GPSR@taylorandfrancis.com
Taylor & Francis Verlag GmbH, Kaufingerstraße 24, 80331 München, Germany

www.ingramcontent.com/pod-product-compliance
Ingram Content Group UK Ltd.
Pitfield, Milton Keynes, MK11 3LW, UK
UKHW050055200926
475702UK00019B/2843
* 9 7 8 0 3 6 7 1 4 3 1 9 0 *